**Cassell Elementary Technical
and Scientific Dictionary**

CASSELL
Elementary Technical and Scientific Dictionary

C. VAUGHAN JAMES
AND
WILLIAM R. LEE

EDITED BY C. VAUGHAN JAMES

Cassell plc
Wellington House
125 Strand
London
WC2R 0BB

387 Park Avenue South
New York
NY 10016–8810

© C. Vaughan James and William R. Lee 1995

All rights reserved. No part of this publication may be reproduced or transmitted in any form or by any means, electronic or mechanical including photocopying, recording or any storage information or retrieval system, without prior permission in writing from the publishers.

First published 1995

British Library Cataloguing-in-Publication Data
A catalogue record for this book is available from the British Library.

ISBN: 0–304–33143–0

Typeset by August Filmsetting, Haydock, St Helens

Printed and bound in Great Britain by MacKays of Chatham

Introduction

The choice of headwords for any dictionary is, of necessity, a matter of individual judgement. Critics tend to comment not so much on what is included as on what has been left out, possibly because the aims of the dictionary and the principles on which it has been structured are not always made clear.

Our aim in selecting items for inclusion has been twofold: to supplement the basic scientific and technological vocabulary of learners whose experience of English has been restricted to 'general' texts, in order to bring their knowledge closer to that of native speakers of English at a comparable level of education; and to introduce in a systematic way a number of terms from various scientific and technological fields in order to facilitate study within those fields. The selection has been made partly on the basis of our own considerable experience of such learners in various parts of the world, and of training and/or advising their teachers, and partly on examination of popular elementary textbooks and authentic texts in common use.

For each word we have provided a definition of its meaning in a scientific or technological context (there may well be other meanings with which we are not here concerned) and, where the definition may not be sufficient, we have added examples of the word in use.

Wherever possible we have indicated after each headword the branch of science or technology in which it is chiefly used. Such a system can obviously be only approximate, since few such words will occur in only one scientific or technological context. We have therefore tried to choose the commonest contexts in which the words may be expected to occur, emphasizing that these are by no means exclusive.

In general, there are three main categories: words which occur most often in a single scientific or technological context; those which occur **only** in such contexts, having been coined to express specific meanings; and finally – the largest group and the one that may present most difficulties – words which are doubtless familiar to the learner in their basic or literary senses but which have been adopted in various scientific or technological texts to convey a meaning of another kind. To the first category belong words such as *atom*, *electron*, *neutron*; to the second, terms like *alternator*, *capacitor*, *systemic*; and to the third, words like *element*, *phase*, *period*.

The choice of 'word fields' (scientific and technological areas of use) has been determined by analysis of the words we thought it necessary to include. Rather than attempting an exhaustive coverage of, say, terms used in Physics (which would be beyond the scope of an 'elementary' dictionary), we first marked, for example, as (*Phys.*), such words in our final selection as would occur most often in texts dealing with Physics. In doing so, we realized that there were branches of Physics (as of other basic scientific or technological areas) to which a considerable number of items belonged, and this led us to list certain of these branches as separate fields or subfields, e.g. Electricity, Radio, Photography, marked as *(Elec.)*, *(Radio)*, *(Photo.)*. It would, of course, have been possible to subdivide **all** the major fields in this way, but this would have resulted in a very large number of minor categories, some of which – for reasons already stated – would in any case have been only approximate.

The complete list of word fields is given below. It will be seen that the broad categories are: Biology, Chemistry, Engineering, Mathematics, Medicine and Physics. Biology includes subdivisions, for our purposes, of Agriculture, Botany, Horticulture and Zoology; Engineering has Aeronautical Engineering, Building, Mechanical Engineering (the biggest category), Metallurgy, Mining and Nautical or Maritime Engineering; Medicine has Psychiatry; Physics has Astronomy, Electricity/Electronics, Optics and Photography; and there are a number of other categories, such as Ecology and Meteorology, which could possibly be assigned to one of the broad categories above, but also may be considered to span several of them.

In square brackets after most items we have provided a phonetic transcription, using symbols from the International Phonetic Alphabet. The system of transliteration is explained on page viii.

Words included in the dictionary are variously described as *a.* (adjective), *n.* (noun), *v.t.* (transitive verb), *v.i.* (intransitive verb), *comb. form* (combining form), etc. It should be understood that many items (only some of which we have indicated) can belong to more than one of these categories. For instance, the noun *contact* is frequently used as a transitive verb *(They contacted the police)* and *fracture* can in one context be a verb and in another context a noun. In such cases, however, the contexts of meaning and syntax will make clear what the precise function of a given item is.

Many scientific and other terms are derived from ancient Greek or from Latin. The names of branches of science, for example, usually end in *-ology* (as in Ecology, Zoology, etc.) and there are many derivatives from each (zoologist, zoological, zoologically, etc.). It would clearly be a waste of space to have listed all of these separately, and we have therefore included only those forms which are irregular or which have a specific and different meaning.

No dictionary is comprehensive, and it is a good idea for students to keep a running list of terms which are **not** included, perhaps following the system of layout that we have used. The golden rule for all learners (and we all go on learning for the whole of our lives) is not to be lazy – USE A DICTIONARY!

Good luck!

C. VAUGHAN JAMES
WILLIAM R. LEE
January 1994

Word Fields

(Aer.) Aeronautics, Aeronautical Engineering
(Agric.) Agriculture
(Astron.) Astronomy
(Biol.) Biology
(Bot.) Botany
(Build.) Building
(Chem.) Chemistry
(Comput.) Computer Science
(Ecol.) Ecology
(Elec.) Electricity, Electronics
(Eng.) Engineering
(Geog.) Geography
(Geol.) Geology
(Hort.) Horticulture
(Maths.) Mathematics
(Meas.) Measurement
(Mech.) Mechanical Engineering
(Med.) Medicine
(Metall.) Metallurgy
(Meteor.) Meteorology
(Mining) Mining Engineering
(Naut.) Nautical Engineering
(Optics) Optics
(Photo.) Photography
(Phys.) Physics
(Psych.) Psychology, Psychiatry
(Radio) Radio
(Zool.) Zoology

Key to Phonetic Symbols

Pronunciation is represented by the following symbols:

Consonant sounds		*Vowel sounds*	
p as in	pot	i as in	tea
b	big	ɪ	pin
t	tin	e	get
d	do	æ	man
k	can	ɑ	father
g	go	ʌ	come
		ɒ	lot
θ	thin	ɔ	all
ð	those	ʊ	put
		u	two
f	four	ɜ	first
v	heavy	ə	about
s	six	ei	day
z	zero	əʊ	no
		ɒʊ	cold, bolt
ʃ	show	aɪ	by
ʒ	vision	aʊ	now
h	hot	ɔɪ	noise
		ɪə	here
m	mix	eə	fair
n	no	ʊə	sure
ŋ	long		
l	oil		
r	red		
j	yellow		
w	wet		

In addition, words consisting of more than one syllable are marked for stress, primary stress being indicated by superscript ' and secondary stress by subscript ˌ, as in [ˈendʒɪn] engine, and [ˌdʒerɒnˈtɒlədʒɪ] gerontology. Short phrases which have a characteristic stress-pattern are also marked for stress.

A single widely-acceptable pronunciation is indicated throughout.

a

A *(abbr.)* atom

abdomen [ˈæbdəmən] *n. (Med.)* the part of the body between the chest and the pelvis in which many internal organs, including the stomach, are situated

abdominal [æbˈdɒmɪn(ə)l] *a. (Med.)* having to do with the abdomen ▶ She complained of severe abdominal pains and was found to have acute appendicitis.

abort[1] [əˈbɔt] *v.t. (Med.)* to terminate a pregnancy early in its development ▶ The danger to the mother's health determined the doctors to abort the foetus.

abort[2] *v.t. (Mech.)* to bring to a premature end ▶ Because of a malfunction of the computer, the launch of the space shuttle was aborted.

abortion [əbˈɔʃ(ə)n] *n. (Med.)* the premature termination of a pregnancy

abrasion [əˈbreɪʒ(ə)n] *n. (Med.)* superficial damage to the skin ▶ Apart from minor cuts and abrasions, the boy who was knocked off his bicycle came to no harm.

abrasive [əˈbreɪsɪv] *n. (Mech.)* a substance used for smoothing or wearing away by friction ▶ Oil and an abrasive cloth will remove most of the rust.

abscess [ˈæbses] *n. (Med.)* a gathering of pus in a tissue or organ ▶ An abscess at the root of a decaying tooth had to be treated with drugs before the dentist would operate.

absolute [ˈæbsəlut] *a. (Chem.)* free from any other substance ▶ Absolute alcohol is alcohol that contains no water.

absolute temperature temperature measured from absolute zero

absolute zero zero ($-273·1°C$) on the absolute scale of temperature

AC *(abbr.)* **alternating current**

accelerator[1] [əkˈseləreɪtə] *n. (Mech.)* a device which increases the flow of fuel into the carburettor of a petrol engine, thus making it run faster ▶ Pressure on the accelerator increases speed.

accelerator[2] *n. (Mech.)* a piece of equipment used in nuclear physics to fire subatomic particles at a target

access [ˈækses] *v.t. (Comput.)* to obtain information from a database ▶ This program will enable you to access data on mainline railway schedules throughout Europe.

accumulate [əˈkjumjəˌleɪt] *v.t.* to amass, store

accumulator [əˈkjumjəˌleɪtə] *n. (Elec.)* an apparatus for storing electrical energy ▶ An accumulator can be recharged by having an electric current passed through it.

acetic [əˈsitɪk] *a. (Chem.)* having to do with the acid that makes vinegar sour

acetone [ˈæsɪtəʊn] *n. (Chem.)* a flammable liquid used as a solvent or in making chloroform

acetylene [ˈəsetəlin] *n. (Chem.)* a gas, composed of carbon and hydrogen, which burns with a brilliant flame ▶ Before the days of easily obtained electricity, acetylene was in common use in lamps to provide light.

acid [ˈæsɪd] *n. (Chem.)* a compound of hydrogen, in which, during a chemical reaction, the hydrogen is replaced by a metal, forming a salt of that metal and water ▶ Strong acids can burn holes in cloth or wood and even dissolve metals.

acid rain rain made acid by such industrial pollutants as sulphur, released into the atmosphere ▶ A large proportion of the Earth's forests are now affected by acid rain.

acoustics[1] [əˈkustɪks] *n. sg. (Phys.)* the scientific study of sound

acoustics² n. pl. (Phys.) the sound properties of a room, etc. ▶ The acoustics of the new concert hall were excellent, and every note could be heard from any part of it.

acquired immune deficiency syndrome see **AIDS**

acre [ˈeɪkə] n. (Meas.) a piece of land containing 4,840 sq.yds. (0·4 ha)

acreage [ˈeɪk(ə)rɪʤ] n. (Meas.) the area of a piece of land, stated in acres ▶ By increasing the acreage of arable land, the farmers were able to grow more crops.

acupuncture [ˈækjʊˌpʌŋktʃə] n. (Med.) a system of treatment in which the surface of the body is punctured by needles at specific points to relieve pain ▶ Acupuncture is a traditional form of medical treatment in China.

acute¹ [əˈkjuːt] a. (Maths.) (of an angle) less than 90°.

acute² a. (Med.) (of an illness, etc.) coming quickly to a climax and therefore requiring urgent treatment ▶ She was found to be suffering from acute appendicitis and had to be operated on without delay.

adaptor [əˈdæptə] n. (Elec.) a device for connecting an electric plug having terminals of one type to a supply point meant for terminals of another type, or for connecting several electrical appliances to one power supply point ▶ When travelling in other countries, it is wise to carry a universal adaptor in your luggage.

addict [ˈædɪkt] n. (Med.: Psych.) a person who is unable to break the habit of using certain drugs, such as heroin, tobacco or alcohol ▶ People who experiment with hard drugs risk becoming addicts.

addiction [əˈdɪkʃ(ə)n] n. (Med.: Psych.) the state or process of being or becoming an addict ▶ Drug addiction is one of the main causes of violent crime.

addictive [əˈdɪktɪv] a. (Med.) (of a drug) which makes the user dependent on it and therefore unable to stop using it ▶ Drugs which seem harmless may in fact be addictive.

addition [əˈdɪʃ(ə)n] n. (Maths.) collecting several numbers or quantities into one

adenoids [ˈædənɔɪdz] n. pl. (Med.) a mass of soft tissue at the back of the nose and throat ▶ Adenoids, when enlarged, impede breathing and give the voice a peculiar quality, unless they are removed surgically.

adhesive [ədˈhiːsɪv] n. or a. a substance used for sticking things together; sticky

adipose [ˈædɪpəʊs] a. (Med.) fat, fatty ▶ Cells in the human body in which fat is stored make up what is called adipose tissue.

adjacent [əˈʤeɪsənt] a. (Maths.) (of two angles) touching, next to each other, as when two lines intersect

adrenal [əˈdriːn(ə)l] a. (Med.) (of a gland) situated near to the kidney

adrenalin [əˈdrenəlɪn] n. (Med.) a hormone secreted by the adrenal gland ▶ An increased flow of adrenalin accompanies a state of excitement.

aeon [ˈiːən] n. (Astron.) an age in the history of the universe; 1,000 million years

aerate [ˈeəreɪt] v.t. (Phys.) to charge a liquid with carbon dioxide ▶ Aerated or 'fizzy' drinks are very refreshing in hot weather.

aerial [ˈeərɪəl] n. (Radio) a device that collects electromagnetic waves for radio or television ▶ Every chimney in the street had a television aerial attached to it.

aerodynamics [ˌeərədaɪˈnæmɪks] n. sg. (Aer.) the science which studies the interaction between air and objects moving through it ▲ Aerodynamics is of fundamental importance in the design of aircraft and missiles.

aerofoil [ˈeərəfɔɪl] n. (Aer.) a wing or tailplane of an aeroplane

aeronautics [ˌeərəˈnɔːtɪks] n. sg. the science of aerial navigation

aerosol [ˈeərəsɒl] n. (Chem.: Mech.) the suspension of fine particles of a substance in air or a gas, or the container from which the suspension can be sprayed ▶ Ecologists are worried about the effect on the Earth's atmosphere of the use of certain aerosols.

aerospace [ˈeərəspeɪs] n. (Phys.: Astron.) the Earth's atmosphere and the space beyond

affinity [əˈfɪnɪtɪ] n. (Chem.) the chemical property that enables certain elements to combine to form new compounds

agent [ˈeɪdʒənt] *n. (Biol.: Chem.)* a substance that precipitates change ▶ The agent of photosynthesis in plants is light.

aggregate [ˈægrɪgət] *n. (Build.)* a mixture of sand and stones used in making concrete

agro- *comb. form* having to do with soil or its cultivation

agrobiology [ˌægrə(ʊ)baɪˈɒlədʒɪ] *n. (Agric.)* the study of plant nutrition in relation to soil management

agrochemical [ˌægrə(ʊ)ˈkemɪk(ə)l] *n. (Agric.)* a chemical for use on the land or in farming, e.g. as a fertilizer

agronomy [əˈgrɒnəmɪ] *n. (Agric.)* the management of land

AIDS (Aids) [eɪdz] *n. (Med.)* Acquired Immune Deficiency Syndrome, a condition in which the immune system of the human body is attacked by a virus which renders it increasingly unable to develop antibodies to combat disease ▶ Persons afflicted with AIDS usually die from such diseases as tuberculosis, which a healthy person can normally resist, or from cancer.

aileron [ˈeɪlərɒn] *n. (Aer.)* the hinged part of the trailing edge of the wing-tip of an aeroplane, used in controlling manoeuvres

airbrake [ˈeəˌbreɪk] *n. (Mech.)* a brake worked by air pressure ▶ Large and powerful motor vehicles are normally equipped with airbrakes.

airbus [ˈeəˌbʌs] *n. (Aer.)* a large, passenger jet aeroplane used for short, intercity flights

air-conditioned [ˈeəkənˌdɪʃ(ə)nd] equipped with a system of circulating cool air ▶ Hotel rooms are usually air-conditioned, especially in hot countries.

air corridor [ˈeəˌkɒrɪdə] an imaginary path, defined in terms of direction and height, to which air traffic is restricted ▶ Adherence to air corridors reduces the risk of mid-air collisions.

air pocket [ˈeəˌpɒkɪt] a downward current of air, which causes an aeroplane to lose height suddenly ▶ Air pockets are most common over stony ground in very hot weather.

air shaft [ˈeəˌʃɑft] a vertical shaft from the surface to the working area of a mine

airframe [ˈeəˌfreɪm] *n. (Eng.)* the structure of a aeroplane, including the engines

airlock[1] [ˈeəˌlɒk] *n. (Mech.)* a stoppage caused by a bubble of air which prevents liquid from passing through a pipe ▶ A bubble of air trapped in the outlet created an airlock and cut off the water supply.

airlock[2] [ˈeəˌlɒk] *n. (Mech.)* a compartment or chamber between two areas with different atmospheric pressures, allowing a person to move from one area to another without loss of pressure, e.g. when entering or leaving a spacecraft

airspeed [ˈeəˌspid] *n. (Aer.)* the speed of an aircraft relative to the air surrounding it ▶ When there is a strong tailwind, the groundspeed is greater than the airspeed.

albino [ælˈbinəʊ] *n. (Med.)* a person who has a congenital absence of pigment in the skin ▶ An albino has white hair and pink eyes.

albumen [ˈælbjʊmɪn] *n. (Chem.: Med.)* a protein, soluble in water, found in the white of eggs and in blood plasma

alcohol [ˈælkəhɒl] *n. (Chem.)* a colourless liquid, formed by fermenting sugar, which is the base of wines and spirits ▶ Alcohol is what makes certain drinks intoxicating.

alcoholism [ˈælkəhɒlɪzəm] *n. (Med.)* addiction to alcohol

algebra [ˈældʒɪbrə] *n. (Maths.)* the branch of mathematics in which letters, symbols and signs are used to express and represent quantities and their interrelationships ▶ Such elementary equations as $x=2$, $y=5$, $x+y=7$ represent a simple form of algebra.

algebraic [ˌældʒɪˈbreɪɪk] *a. (Maths.)* having to do with algebra ▶ The problem was formulated as an algebraic equation.

alimentary [ˌælɪˈment(ə)rɪ] *a. (Biol.: Med.)* having to do with nutrition

alimentary canal the tube in the human body into which food is taken, via the mouth, and from which waste matter is excreted, via the anus

alkali [ˈælkəlaɪ] *n. (Chem.)* a compound of hydrogen and oxygen with other substances, which is soluble in water and can neutralize acids

alkaloid [ˈælkəlɔɪd] *n.* (*Chem.*) one of the nitrogen-based substances derived from plants and sometimes used as medicinal drugs

allergic (to) [[əˈlɜdʒɪk] *a.* (*Med.*) having an allergy ▶ Bee stings can be fatal to people who are allergic to them. ▶ Before administering drugs, doctors have to make sure that the patients are not allergic to them.

allergy [ˈælədʒɪ] *n.* (*Med.*) hypersensitivity to certain substances, especially foods and drugs, which are perfectly acceptable to most people

alleviate [əˈliːvɪeɪt] *v.t.* (*Med.*) to lessen, lighten ▶ The doctor prescribed a mixture to alleviate the pain.

alloy [ˈælɔɪ] *n.* (*Metall.*) a mixture of metals of different values ▶ Bronze is an alloy of tin and copper.

alluvial [əˈljuːvɪəl] *a.* (*Geol.*) (of soil) deposited by rivers and floods ▶ The alluvial soil of the Nile valley is extremely fertile.

alternate [ˈɔltəneɪt] (v.) *v.i.* (*Elec.*) to change from a positive to a negative electric current and back again

alternating current [ˈɔltəneɪtɪŋ ˌkʌrənt] an electric current that changes direction at regular intervals

alternator [ˈɔltəneɪtə] *n.* (*Elec.: Mech.*) a dynamo for generating an alternating electric current ▶ Car engines have alternators in order to keep their batteries charged.

altimeter [ˈæltɪmɪtə] *n.* (*Phys.*) an instrument used for measuring height above a certain level, usually sea-level ▶ Many air accidents have been caused by faulty altimeters.

altitude [ˈæltɪtjud] *n.* (*Phys.: Geog.*) height above sea-level ▶ Breathing is difficult at high altitudes due to the rarefied atmosphere. ▶ The altitude of the peaks is given on the map.

aluminium [ˌæljəˈmɪnjəm] *n.* (*Metall.*) a white, easily-worked metallic element, with a high resistance to corrosion ▶ Aluminium is used in making various alloys which are useful in aircraft construction.

amalgam [əˈmælgəm] *n.* (*Metall.*) a mixture of any other metal with mercury ▶ Certain amalgams are used in dentistry for filling teeth.

amnesia [æmˈniːzɪə] *n.* (*Med.*) loss of memory ▶ Head injuries may result in temporary amnesia.

amoeba [əˈmiːbə] *n.* (*Biol.*) a living organism which consists of a single cell

amorphous [əˈmɔːfəs] *a.* (*Biol.*) having no fixed shape, or an abnormal shape ▶ An amoeba is an amorphous organism, which constantly changes shape.

ampere [ˈæmpeə] (*abbr.* **amp**) *n.* (*Elec.*) a unit by which an electrical current is measured ▶ 1 ampere is the current sent by 1 volt through a resistance of 1 ohm.

amphetamine [æmˈfetəmɪn] *n.* (*Chem.: Med.*) a derivative of a synthetic drug which stimulates the brain ▶ Perhaps the best known amphetamine is 'Speed'.

amphibian [æmˈfɪbɪən] *n.* (*Zool.*) an animal which can live both on land and in water ▶ Probably the most commonly known amphibian is the frog.

amphibious [æmˈfɪbɪəs] *a.* (*Zool.*) (of an animal) able to live on land or in water

amplifier [ˈæmplɪˌfaɪə] *n.* (*Phys.: Mech.*) a piece of apparatus which makes sounds louder ▶ Such instruments as an electric guitar are relayed through an amplifier.

amplitude [ˈæmplɪtjud] *n.* (*Phys.*) the distance between the top and bottom of a sound wave

ampoule [ˈæmpul] *n.* (*Med.*) a sealed phial containing one dose of a drug

amputate [ˈæmpjuˌteɪt] *v.t.* (*Med.*) to cut off a limb by surgery ▶ The crash victim's hopelessly crushed leg was amputated at the knee.

anaemia [əˈniːmɪə] *n.* (*Med.*) a medical condition caused by a shortage of red cells in the blood, or by loss of blood

anaesthesia [ˌænɪsˈθiːzɪə] *n.* (*Med.*) the condition of being unable to feel pain, so that a surgical operation may be carried out without suffering

anaesthetic [ˌænɪsˈθetɪk] *n.* (*Med.*) a substance which produces loss of sensitivity to pain

anaesthetist [əˈnisθətɪst] *n.* (*Med.*) someone qualified to administer an anaesthetic

anatomy [əˈnætəmɪ] *n.* (*Biol.: Med.*) the physical structure of animals and plants, or the study of this ▶ The study of human anatomy is an important element in medical training.

animate [ˈænɪmət] *a.* (*Biol.*) living

anneal [əˈnil] *v.t.* (*Metall.*) to temper a metal by first heating it, then cooling it slowly ▶ Annealing a metal reduces internal stresses and renders it less likely to break.

annular [ˈænjʊlə] *a.* (*Astron.*) ring-shaped ▶ In an annular eclipse, the moon obscures the centre of the sun only, leaving a ring of light around it.

anode [ˈænəʊd] *n.* (*Elec.*) the positive electrode

anorexia [ˌænəˈreksɪə] *n.* (*Med.*) loss of appetite

anorexia nervosa a psychological disorder characterized by an aversion to eating and fear of gaining weight

ante- *comb. form* before

antenatal [ˌæntɪˈneɪt(ə)l] *a.* (*Med.*) before birth, during the period of pregnancy ▶ Regular attendance at an antenatal clinic is strongly advised for all pregnant women.

antenna¹ [ænˈtenə] *pl.* **antennae** [ænˈteni] *n.* (*Biol.*) one of a pair of sensitive feelers on the head of an insect, which enable it to detect food etc.

antenna² *n.* (*Radio*) an aerial which picks up radio signals

anthropoid [ˈænθrəpɔɪd] *a.* (*Biol.*) having a form resembling a human being

anthropology [ˌænθrəˈpɒlədʒɪ] *n.* the scientific study of mankind in all its aspects

anti- *comb. form* opposite, opposed to, against, instead of

antibiotic [ˌæntɪbaɪˈɒtɪk] *n.* (*Biol.: Med.*) a substance produced by one micro-organism which inhibits growth of, or kills, another micro-organism ▶ The development of antibiotics like penicillin has saved countless lives.

antibody [ˈæntɪbɒdɪ] *n.* (*Biol.*) a substance developed in the body in response to the presence of a toxin ▶ The purpose of inoculation against disease is to stimulate the development of antibodies to counter it.

anticoagulant [ˌæntɪkə(ʊ)ˈægjʊl(ə)nt] *n. or a.* (*Chem.: Med.*) a drug that hinders blood-clotting ▶ Aspirin is regularly taken as an anticoagulant by sufferers from mild heart disease.

anticyclone [ˌæntɪˈsaɪkləʊn] *n.* (*Meteor.*) an outward movement of air from an area of high pressure ▶ The movement of air from a cyclone is in a clockwise direction, but from an anticyclone the air moves anti-clockwise.

antidote [ˈæntɪdəʊt] *n.* (*Med.*) a medicine designed to counteract a specific poison or disease ▶ Antidotes are now available for most known diseases. ▶ Some snake bites are fatal unless the correct antidote is quickly available.

antiseptic [ˌæntɪˈseptɪk] *n.* (*Med.*) a substance that counteracts the development of sepsis ▶ The wound was washed with an antiseptic to prevent the development of bacterial infection.

anus [ˈeɪnəs] *n.* (*Biol.*) the lower end of the passage in animals and humans through which waste matter is excreted

aorta [eɪˈɔtə] *n.* (*Biol.*) the largest artery in the body, carrying blood from the heart to the rest of the body

aperient [əˈpɪərɪənt] *n.* (*Med.*) a laxative medicine

aperture [ˈæpətʃə] *n.* (*Mech.*) a small hole or cavity ▶ The variable hole through which the light passes when a film is exposed in a camera is called the aperture.

apex [ˈeɪpeks] *n.* (*Maths.*) (of a triangle) the highest point

aphasia [əˈfeɪzɪə] *n.* (*Med.*) loss or partial loss of the ability to communicate by language

apogee [ˈæpədʒi] *n.* (*Astron.*) the point in the orbit of the Moon or any planet or satellite at which it is at the greatest distance from the Earth

apparatus [ˌæpəˈreɪtəs] *n.* (*Mech.*) a piece of

equipment ▶ The chemistry laboratory is equipped with the latest apparatus for conducting the experiments.

appendectomy [ˌæpenˈdektəmɪ] *n.* *(Med.)* removal of the appendix by surgery

appendicitis [əˌpendɪˈsaɪtɪs] *n.* *(Med.)* inflammation of the appendix

appendix [əˈpendɪks] *n.* *(Biol.: Med.)* a small and narrow closed branch of the intestine

appliance [əˈplaɪəns] *n.* *(Mech.)* a device or apparatus for performing a specific task

aqueous [ˈeɪkwɪəs] *a.* *(Chem.: Geol.)* consisting of, containing, or deposited by water

arable [ˈærəbl] *a.* *(Agric.)* (of land) in which crops are or may be grown ▶ The farmland was divided into grazing for cattle and arable land for wheat.

arc¹ [ɑk] *n.* *(Maths.)* part of the circumference of a circle

arc² *n.* *(Elec.)* the luminous arched bridge between two electrodes when a current is passed between them

arc-lamp an electric lamp in which the source of light is the arc between the two electrodes

arc-weld to weld metal by means of an electric arc

archaeology [ˌɑkiˈɒlədʒɪ] *n.* the scientific study of Man's past history, especially through the excavation of ancient remains

archipelago [ˌɑkiˈpeləgəʊ] *n.* *(Geog.: Geol.)* an area of sea studded with islands ▶ The original Greek term 'archipelago' was used to describe the Aegean sea.

arid [ˈærɪd] *a.* *(Geog.: Geol.)* dry, barren, lacking in moisture ▶ In the arid zone or desert, very little can be grown.

aridity [əˈrɪdətɪ] *n.* *(Geog.: Geol.)* (of soil) the state of being dry or parched

arithmetic [əˈrɪθmətɪk] *n.* *(Maths.)* the study of numbers and calculations

arithmetical [ˌærɪθˈmetɪk(ə)l] *a.* *(Maths.)* having to do with arithmetic ▶ The four basic forms of arithmetical calculation are addition, subtraction, multiplication and division.

arithmetical progression a series of numbers that increase or decrease by the same amount ▶ The series 2−4−6−8, and 8−6−4−2, are arithmetical progressions.

armature [ˈɑmətʃə] *n.* *(Elec.: Mech.)* the revolving part of an electric motor or dynamo

artery [ˈɑtərɪ] *n.* *(Biol.: Med.)* one of the blood vessels conveying oxygenated blood from the heart to other parts of the body

arthritis [ɑˈθraɪtɪs] *n.* *(Med.)* inflammation of the joints

articulate [ɑˈtɪkjʊˌleɪt] *v.t.* *(Mech.)* to connect or join together

articulated lorry a lorry with a trailer connected so that it can turn at an angle ▶ The articulated lorry had jack-knifed and blocked the road.

ascendant [əˈsendənt] *a.* *(Astron.)* (of a star etc.) rising, moving toward its highest point

ascorbic [əˈskɔbɪk] *a.* *(Chem.: Med.)* (of an acid) relating to vitamin C, as found in fruit and vegetables

asexual [eɪˈseksjʊəl] *a.* *(Biol.)* without active sexual organs

asphalt [ˈæsfælt] *n.* *(Build.)* a dark, bituminous limestone mixed with tar and used for roofing or road-surfacing

asphyxia [əˈsfɪksɪə] *n.* *(Med.)* death due to failure to inhale oxygen ▶ Most fire victims die not from burns, but from asphyxia caused by smoke.

aspirate [ˈaespəreɪt] *v.t.* *(Med.)* to remove body fluids by suction ▶ The superfluous fluid in the lungs was aspirated with a large syringe.

aspirin [ˈæsp(ə)rɪn] *n.* *(Med.)* a mild drug, commonly prescribed as a pain-killer

assemble [əˈsemb(ə)l] *v.t.* *(Mech.)* to fit together the various parts of an apparatus ▶ The laboratory technicians soon assembled the newly arrived apparatus.

assembly line [əˈsemblɪ ˌlaɪn] a line of workers and/or machines operating various stages in the assembling of an industrial product ▶ Work on an assembly line is very repetitive.

asteroid [ˈæst(ə)rɔɪd] *n.* *(Astron.)* any of the small heavenly bodies that orbit the sun

asthma [ˈæsmə] *n.* *(Med.)* a breathing disorder

characterized by wheezing, constriction of the chest, and coughing ▶ Asthma can be greatly relieved by use of a nebulizer.

asthmatic [æsˈmætɪk] *a.* or *n.* *(Med.)* having to do with, or suffering from, asthma, or a person who suffers from asthma

astronaut [ˈæstrənɔt] *n.* *(Aer.)* a person who travels in space beyond the Earth's atmosphere in a spaceship, etc.

astronomy [əˈstrɒnəmɪ] *n.* the scientific study of heavenly bodies

astrophysics [ˌæstrəʊˈfɪzɪks] *n. sg.* the study of the physics of heavenly bodies

atmosphere [ˈætməˌsfɪə] *n.* *(Phys.)* the layer of gases surrounding the Earth or other heavenly bodies

atmospheric pressure a unit of pressure calculated as the average pressure of the Earth's atmosphere at sea-level ▶ Average atmospheric pressure is taken as equal to about 15 pounds to the square inch (15 lb/sq. in.).

atmospherics [ˌætməˈsferɪks] *n. pl.* *(Radio)* radio interference caused by the presence of electricity in the atmosphere

atom [ˈætəm] *n.* *(Phys.: Chem.)* the smallest possible particle of an element ▶ An atom consists of a nucleus surrounded by electrons.

atomic [əˈtɒmɪk] *a.* *(Phys.: Chem.)* consisting of, or having to do with, atoms

atomic bomb a bomb in which the explosion is due to atomic energy, released by the nuclear fission of uranium, plutonium etc.

atomic energy the energy released when the nucleus of an atom changes by fission (uranium, plutonium) or fusion (hydrogen) ▶ The main peaceful application of atomic energy is in the generation of electricity.

atomic pile [əˌtɒmɪk ˈpaɪl] a nuclear reactor

atomic weight [əˌtɒmɪk ˈweɪt] the weight of an atom of a substance compared with that of an atom of carbon

atomizer [ˈætəmaɪzə] *n.* *(Mech.)* an instrument for reducing a liquid, such as a perfume or a disinfectant, into a spray

atrophy [ˈætrəfɪ] *n.* *(Med.)* wasting of the body or an organ from lack of nutrition or use

audible [ˈɔdəb(ə)l] *a.* *(Phys.)* (of a sound) loud enough to be heard ▶ The sound of an approaching aircraft was clearly audible in the distance.

audible range the range of loudness within which a sound can be heard ▶ The aircraft moved quickly away and was soon out of audible range.

audio- *comb. form* having to do with hearing or the reproduction of sound

audiovisual [ˌɔdɪəʊˈvɪʒʊəl] *a.* *(Phys.)* directed at or involving sound and sight ▶ Teachers now make extensive use of audiovisual aids, such as films, tape recordings and video.

auditory [ˈɔdɪt(ə)rɪ] *a.* *(Biol.: Phys.)* concerned with hearing ▶ Auditory skills are very important in the study of languages and music.

autism [ˈɔtɪzm] *n.* *(Med.: Psych.)* a mental state marked by an inability to speak to other people or to establish personal relationships

autistic [ɔˈtɪstɪk] *a.* *(Med.: Psych.)* suffering from autism

auto- *comb. form* independently

automated [ˈɔtəmætɪd] *a.* *(Mech.)* made automatic ▶ All the assembly lines in the new car factory are fully automated.

automatic[1] [ɔtəˈmætɪk] *a.* *(Mech.)* self-regulating, acting on its own, without external aid ▶ All the locks on the doors of the building are fully automatic.

automatic[2] *n.* or *a.* *(Mech.)* a gun which continues to fire as long as the trigger is depressed ▶ An automatic pistol fires a number of rounds in a very short time.

automatic pilot a device which automatically maintains an aircraft or spacecraft on a previously-determined course without the manual participation of the crew

automatic transmission (in a motor vehicle) automatic changing of gears without manual participation by the driver

automation [ɔtəˈmeɪʃ(ə)n] *n.* *(Eng.)* the use of self-regulating or automatically programmed machines to perform functions in a manufacturing process formerly car-

autopsy | 8 | **axis**

ried out by people ▶ The automation of industrial processes leads to increased productivity and the employment of fewer people.
autopsy [ˈɔtɒpsɪ] *n.* (*Med.*) the medical examination of a body to establish the cause of death ▶ Since the cause of death was unclear, the coroner requested an autopsy.
AV (*abbr.*) **audiovisual**
aviation [ˌeɪvɪˈeɪʃ(ə)n] *n.* (*Aer.: Eng.*) the design and manufacture of aircraft
avionics [ˌeɪvɪˈɒnɪks] *n. sg.* (*Radio: Elec.*) the science of the development and use of electrical and electronic equipment in aircraft and spacecraft

axiom [ˈæksɪəm] *n.* (*Maths.*) a self-evident proposition that does not require proof
axiomatic [ˌæksɪəˈmætɪk] *a.* (*Maths.*) (of a proposition) self-evidently true
axis [ˈæksɪs] *n.* (*Maths.*) a real or imaginary straight line around which a body revolves ▶ The earth revolves about an axis running through its centre between the two poles.
axle [ˈæksl] *n.* (*Mech.*) the pin or bar on which a wheel revolves ▶ The back axle of the lorry snapped on the rough track and the wheels would no longer turn.

b

bacillus [bəˈsɪləs] *n. (Biol.)* an extremely small living organism, a type of bacterium, often associated with disease

bacteria [bækˈtɪərɪə] *n. pl. (sg.* **bacterium** [bækˈtɪərɪəm]) *(Biol.)* tiny, single-cell organisms present in air, water, and soil, and in all creatures, living or dead

bacteriology [bækˌtɪərɪˈɒlədʒɪ] *n. (Biol.)* the study of bacteria

baffle [ˈbæfl] *v.t. (Mech.)* to regulate the flow of a liquid or a noise

baffle-board a device to prevent the carrying of noise

baffle-plate a plate used to direct the flow of a fluid

balance [ˈbæl(ə)ns] *n. (Mech.)* a pair of scales or other instrument used for measuring the weight of small quantities of substances

ball and socket [ˌbɔː l(ə)n ˈsɒkɪt] the working part of a joint composed of a ball resting in a socket, which enables movement in any direction

ball-bearing [ˈbɔːlˌbeərɪŋ] one of a number of loose metallic balls used in bearings to lessen friction between moving parts of a machine

ballcock [ˈbɔːlkɒk] *n. (Mech.)* a hollow ball at the end of a lever, which acts as a valve by rising or falling with the level of the water in a cistern, admitting or cutting off the flow of water by raising or lowering the other end of the lever ▶ Most lavatory cisterns are controlled by simple ballcocks.

ballistic missile a military missile guided over the first part of its course and then descending according to the laws of ballistics ▶ The mere existence of intercontinental ballistic missiles preserved the balance of terror for almost half a century.

ballistics [bəˈlɪstɪks] *n. sg. (Mech.)* the scientific study of the movement of objects projected into the air ▶ Ballistics experts can often identify the gun from which a bullet has been fired.

band [bænd] *n. (Radio)* the range of frequencies which is used for a particular radio transmission, or within which a piece of equipment is most efficient ▶ Bands are usually described in terms of metres.

bank¹ [bæŋk] *v.i. (Aer.)* (of an aircraft) to change direction by raising one wing and operating the rudder ▶ The low-flying aircraft banked steeply to avoid the tall pylons.

bank² *n. (Elec.)* a number of pieces of equipment linked together ▶ The wall of the studio was lined with a bank of television monitors.

bar see **millibar**

bar code [ˈbɑː kəʊd] a compact arrangement of lines of various lengths and thicknesses containing information such as price, which is printed on the packaging of goods for sale, and which can be read by an electronic device at the point of purchase

barbiturate [bɑːˈbɪtʃʊrət] *n. (Chem.: Med.)* a drug that has a sedative effect ▶ Many people take mild doses of barbiturates as sleeping pills, but improperly used they can become dangerously addictive.

barometer [bəˈrɒmɪtə] *n. (Phys.: Meteor.)* an instrument which measures atmospheric pressure ▶ When the pressure indicated by the barometer falls, bad weather is probably on the way.

barrage [ˈbærɑːʒ] *n. (Eng.)* an artificial bar or dam placed across a river to control the level of the water

base¹ [beɪs] *n. (Chem.) a* substance with which an acid combines to form a salt

base² *n. (Maths.)* (of a triangle) the lowest side

base³ *n. (Maths.)* the number on which a system of calculations depends ▶ The traditional use of base 10 for calculations was replaced in 'new maths' by the binary system, using base 2.

basin [ˈbeɪs(ə)n] *n. (Geog.)* the area of land which drains into a river ▶ The Amazon basin is still largely covered by rainforest.

bathysphere [ˈbæθɪsfɪə] *n. (Eng.: Biol.)* a large steel vessel, able to resist high pressures, which is lowered into deep water to facilitate observation of marine life

battery [ˈbætərɪ] *n. (Elec.)* a series of electric cells, etc., which generate or store electrical energy ▶ Without a battery to provide the spark, a petrol engine will not start.

bearing [ˈbeərɪŋ] *n. (Mech.)* any part of a machine that bears the friction from moving parts ▶ The old engine became extremely noisy because the bearings were very worn.

bellows [ˈbeləʊz] *n. pl. (Mech.)* a device that creates a flow of air and directs it in a specific direction ▶ Giant bellows were used in the old iron foundry to keep the furnace burning.

benign [bɪˈnaɪn] *a. (Med.)* (of a tumour etc.) not threatening to life ▶ Fears that the tumour was malignant were dispelled when medical tests revealed that it was in fact benign.

berth¹ [bɜθ] *v.t. and v.i. (Naut.)* to manoeuvre a ship into place at a quay ▶ The sea was so rough that the liner could not be berthed and had to drop anchor in the harbour.

berth² *n. (Naut.)* a place at a quay where a ship is tied up when not in use

bevel [ˈbevl] *v.t. (Mech.)* to shape the edge of a piece of metal or wood at an angle ▶ The corners of the furniture were bevelled in order to lessen the possibility of injury.

bi- *comb. form* double, doubly

bifocal [baɪˈfəʊkl] *a. (Optics)* having two foci ▶ Bifocal spectacles have two foci, one for near vision and one for distant vision.

bifurcate [ˈbaɪfəkeɪt] *v.i. (Maths.)* to divide into two branches

bilateral [baɪˈlæt(ə)r(ə)l] *a. (Biol.)* having two sides ▶ Vertebrate creatures are bilateral in structure, so that one half is almost a mirror image of the other.

bile [baɪl] *n. (Biol.)* a bitter, yellowish fluid secreted by the liver to aid digestion

binary [ˈbaɪnərɪ] *a. (Maths.)* consisting of a pair or pairs

binary compound a chemical compound of two elements

binary notation a number system using base 2 instead of base 10 ▶ In the base 2 system used in computers, 1 and 0 can represent the two states, on–off.

binomial [baɪˈnəʊmɪəl] *a.* or *n. (Maths.)* a mathematical expression consisting of two terms, united by the signs + or −

bio- *comb. form* having to do with life

biochemistry [ˌbaɪəʊˈkemɪstrɪ] *n. (Chem.)* the scientific study of the chemical processes that take place in living things

biodegradable [ˌbaɪəʊdɪˈgreɪdəbl] *a. (Biol.: Chem.)* capable of being broken down by bacteria ▶ Ecologists encourage greater use of biodegradable materials instead of virtually indestructible plastic for packaging manufactured products.

biology [baɪˈɒlədʒɪ] *n.* the science of living matter in all its forms

biophysics [ˌbaɪəʊˈfɪzɪks] *n. sg.* the science of the application of physics to living things

biopsy [ˈbaɪɒpsɪ] *n. (Med.)* the removal of living tissue for microscopic examination and diagnosis ▶ The doctor conducted a biopsy on the patient's glands before deciding whether to operate.

biotechnology [ˌbaɪəʊtekˈnɒlədʒɪ] *n. (Biol.: Mech.)* the use of micro-organisms and biological processes in industry

bit¹ [bɪt] *n. (Comput.)* a unit of information in computers and information theory, representing one of the two states – on–off. (Binary digit)

bit² *(Mech.)* the revolving part of a drill ▶

The carpenter drilled a series of holes with a brace and bit.

bitumen [ˈbɪtʃʊmɪn] *n. (Chem.)* one of the sticky mixtures of hydrocarbons that are produced as residue from the distillation of petroleum ▶ Bitumen is used in surfacing roads.

bituminous [bɪˈtʃuːmɪnəs] *a. (Build.)* containing bitumen

bivalent [baɪˈveɪlənt] *a. (Chem.)* (of an element) having a chemical valency of two

bivalve [ˈbaɪvælv] *n. (Biol.)* a creature which has two shells or valves, which open and shut ▶ Oysters and other shellfish are bivalves.

bladder [ˈblædə] *n. (Med.)* the organ of the body in which urine collects before release

blade [bleɪd] *n. (Mech.)* (of a fan etc.) a broad surface of metal, plastic or wood, set at an angle in order to deflect the flow of air or water in a specific direction ▶ As the fan in the ceiling revolved, its blades distributed a stream of air throughout the room.

blast[1] [blɑːst] *v.t. (Mining)* to break up rocks with an explosive charge ▶ The red flag indicates that rocks are being blasted in the quarry.

blast[2] *v.t. (Eng.)* to direct a strong current of air, e.g. with a fan or bellows

blast furnace [ˈblɑːst ˌfɜːnɪs] an iron-smelting works in which the furnace is maintained by a continuous blast of air directed on to it

blast-off the launch of a space vehicle or missile

bleach [bliːtʃ] *v.t. (Chem.)* to make something white through the use of chemical agents

bleaching-powder chloride of lime

bleep[1] [bliːp] *v.t. (Elec.)* to summon with the use of an electronic bleeper

bleep[2] *n. (Elec.)* the sound made by a bleeper, or the bleeper itself

bleeper [ˈbliːpə] *n. (Elec.)* an electronic device, of the sort frequently carried by doctors, which is used to summon them by emitting a bleeping sound

blight[1] [blaɪt] *n. (Bot.)* a disease in plants, caused by fungi or various kinds of insects ▶ Crops may be sprayed with fungicides and insecticides to avoid blight.

blight[2] *v.t. (Bot.)* to infect with blight ▶ The entire crop of apples was blighted and had to be destroyed.

blip [blɪp] *n. (Radio)* an abnormal trace on a radar screen indicating the presence of a vessel, aircraft or other body ▶ A series of blips on the screen betrayed the approach of several aircraft.

block and tackle [ˌblɒk (ə)n ˈtæk(ə)l] a system of pulleys mounted on a frame and used for lifting heavy objects ▶ The car was lifted out of the ditch with a block and tackle.

blood [blʌd] *n. (Biol.)* the red fluid circulating via arteries and veins through the bodies of vertebrate animals

blood bank [ˈblʌd ˌbæŋk] a store where supplies of blood for transfusions are kept

blood count [ˈblʌd ˌkaʊnt] a calculation of the number of red and white corpuscles in a sample of blood

blood donor [ˈblʌd ˌdəʊnə] a person who donates blood to a blood bank

blood group [ˈblʌd ˌɡruːp] one of the four groups into which human blood has been classified, so that compatible blood may be used in transfusions

blood plasma [ˈblʌd ˌplæzmə] blood from which all the red corpuscles have been removed

blood pressure [ˈblʌd ˌpreʃə] the pressure of the blood on the walls of the arteries

blood transfusion [ˈblʌd trænˈsfjuːʒ(ə)n] the transference of blood from one person or animal to another ▶ The injured driver had lost a lot of blood and was given an immediate transfusion.

blood vessel ˈblʌd ˌves(ə)l] an artery or vein, through which the blood circulates in the body

bloodstream [ˈblʌd ˌstriːm] the flow of blood circulating through the body

blowlamp [ˈbləʊˌlæmp] *n. (Mech.)* a torch that produces a jet of flame and may be used for burning off paint etc. (*also called* **blow torch**)

blueprint [ˈbluːˌprɪnt] *n. (Eng.)* a basic plan,

drawn in white on blue paper, showing designs for building construction, electrical circuits etc.

bodily ['bɒdɪlɪ] *a.* having to do with a body

body ['bɒdɪ] *n.* any object that has definable size, composition, etc.

boil[1] [bɔɪl] *n.* (*Med.*) a hard, inflamed swelling on a living body ▶ A painful boil on the patient's arm had to be lanced and cleaned before it could heal.

boil[2] *v.t.* and *v.i.* (*Phys.*) to heat a liquid to the temperature at which it vaporizes

boiler ['bɔɪlə] *n.* (*Eng.*) a container in which water is boiled ▶ The boiler of the old steam locomotive built up a great head of steam.

boiling point ['bɔɪlɪŋˌpɔɪnt] the temperature at which a liquid turns into a gas ▶ The boiling-point of water at sea-level is 100°C (212°F).

bollard ['bɒlɑd] *n.* (*Naut.*) a large post on a wharf, etc., to which ships may be tied, or in a street to prevent traffic from passing

bolt[1] [bɒʊlt] *n.* (*Mech.*) a metal pin, usually with a screw at one end to receive a nut, used for holding objects together, especially pieces of metal ▶ The various parts of the apparatus were held firmly together with a series of nuts and bolts.

bolt[2] *v.t.* (*Mech.*) to fasten with bolts ▶ The heavy machines were bolted to the deck so that they could not move when the ship pitched or rolled.

bonnet ['bɒnɪt] *n.* (*Eng.*) (of a car) the hinged metal panel covering the engine

boom [bum] *n.* (*Eng.*) a movable chain, net or other obstruction placed at the mouth of a harbour to prevent the passage of vessels

boost [bust] *v.t.* (*Elec.: Mech.: Med.*) to increase the strength of something, e.g. to raise the voltage of electric current in a circuit, raise the pressure in an internal combustion engine, or supplement the dose of an injection ▶ The cholera injection was repeated after ten days in order to boost its strength.

booster ['bustə] (*Elec.: Mech.: Med.*) something that increases strength or power ▶ The electric heater booster switches the current on automatically at midday. ▶ The booster rocket was fired to take the satellite out of the Earth's gravity into orbit.

boot [but] *n.* (*Mech.*) the luggage compartment

bore [bɔ] *n.* (*Geog.*) a high wave, caused by movements of the tides, which runs quickly up the narrow estuary of a river

boss [bɒs] *n.* (*Mech.*) an enlarged section of a shaft, such as a crankshaft

botanic gardens an area devoted to the scientific cultivation, study and display of plants ▶ The botanic gardens in Kew have a huge collection of exotic plants.

botany ['bɒtənɪ] *n.* (*Biol.*) the scientific study of plants

botulism ['bɒtjʊlɪz(ə)m] *n.* (*Med.*) a form of food poisoning caused by eating preserved food infected with a botulus bacillus

bowel ['baʊəl] *n.* usually *pl.* (*Med.*) the intestines, gut

brace [breɪs] *n.* (*Mech.*) a strap or length of wood or metal used to tie things together

brace and bit [ˌbreɪs (ə)n 'bɪt] a tool used for making holes, consisting of a crank (brace) and a revolving drill (bit)

breed [brid] *v.t.* and *v.i.* (*Agric.: Biol.*) to raise animals in order to produce young, or (of animals) to produce young

breeder reactor ['bridə rɪˌæktə] a nuclear reactor which produces more plutonium than it consumes

breeze [briz] *n.* (*Build.*) coal and coke ashes used to make blocks for building

breeze block ['briz ˌblɒk] a large brick or block made of cinders and cement

bromide ['brəʊmaɪd] *n.* (*Chem.: Med.*) a kind of sedative containing bromine

bronchi ['brɒŋkaɪ] *n. pl.* (*sg.* **bronchus**) ['brɒŋkəs] (*Med.*) the main divisions of the windpipe, and their branches in the lungs

bronchial ['brɒŋkɪəl] *a.* (*Med.*) having to do with the breathing passages (bronchi) ▶ A common cause of death in old age in cold countries is bronchial pneumonia.

bronchitis [brɒŋˈkaɪtɪs] *n. (Med.)* inflammation of the bronchi ▶ Smoking and air pollution are contributory causes of acute bronchitis.

bronze [brɒnz] *n. (Metall.)* an alloy of copper and tin with a distinctive brown colour

buffer [ˈbʌfə] *n. (Mech.)* a device for deadening the shock of a collision between two objects ▶ Fortunately, there was no damage from the impact when the train hit the buffers.

bulkhead [ˈbʌlkhed] *n. (Eng.)* a partition which divides one part of a ship or aircraft from another

bulwark [ˈbʊlwək] *n. (Eng.)* the part of the side of a ship that rises above the upper deck

bunker [ˈbʌŋkə] *n. (Mech.)* a large container for holding solid fuels, such as coal

Bunsen burner [ˈbʌns(ə)n ˌbɜnə] a lamp in which a mixture of gas and air produces an intense flame, used for heating vessels in chemistry laboratories ▶ The Bunsen burner is named after its inventor, Robert Wilhelm Bunsen (1811–99).

buoy [bɔɪ] *n. (Mech.)* an anchored float used to mark a fairway or an obstruction at sea

buoyancy [ˈbɔɪənsɪ] *n. (Phys.)* the ability to float, loss of weight from immersion in a liquid

bush [bʊʃ] *n. (Mech.)* the metal lining of an axle hole or similar aperture

butane [ˈbjuteɪn] *n. (Chem.)* a flammable gas of the paraffin type, derived from petroleum ▶ Bottled butane is used for cooking when there is no mains fuel supply.

bypass [ˈbaɪpəs] *n. (Med.)* a surgical operation to ensure a flow of arterial blood by bypassing blocked or damaged arteries ▶ Many heart-disease patients have been given a new lease of life through bypass operations.

byte [baɪt] *n. (Comput.)* a series of usually eight binary digits, treated as a unit

C

cable [ˈkeɪbl] *n.* (*Mech.: Elec.*) a strong rope of hemp or steel, or an electrical circuit holding more than one conductor in the same insulated cover ▶ The ship was tied up at the quay with thick, hemp cables at each end. ▶ The high-tension electricity was carried in rubber-covered cables on a line of pylons.

cable railway a funicular railway

cable television a system of supplying TV programmes to individual houses etc. by underground cable

cadaver [kəˈdævə] *n.* (*Med.*) a dead body, corpse

Caesarian section [siˌzeərɪən ˈsekʃ(ə)n] the delivery of a baby through the walls of the abdomen ▶ Julius Caesar was said to have been born in the manner now named after him.

caffeine [ˈkæfin] *n.* (*Chem.*) a vegetable alkaloid derived from coffee and tea, which acts as a mild stimulant ▶ Some people drink only decaffeinated coffee at night because they believe that caffeine prevents them from sleeping.

cage [keɪdʒ] *n.* (*Mining*) a cage-like lift in which miners travel between the surface and the underground workings of a mine

calcium [ˈkælsɪəm] *n.* (*Chem.*) a silver-white metallic element

calcium chloride bleaching powder

calcium oxide lime

calculate [ˈkælkjʊˌleɪt] *v.i.* and *v.t.* (*Maths.*) to determine by a mathematical process

calculation [ˌkælkjʊˈleɪʃ(ə)n] *n.* (*Maths.*) the process of calculating or the result obtained

calculator [ˈkælkjʊˌleɪtə] *n.* (*Elec.*) an electronic device, usually small and portable, which can carry out mathematical calculations

calculus [ˈkælkjʊləs] *n.* (*Maths.*) a branch of mathematics dealing with variable quantities ▶ Differential calculus may be concerned with measuring the speed at which an object is moving, and integral calculus with the distance it has travelled.

calibrate [ˈkælɪbreɪt] *v.t.* (*Maths.*) to mark a scale on a measuring instrument, such as a thermometer or scales ▶ A thermometer is calibrated by taking the level of mercury at freezing- and boiling-points and subdividing the distance between them.

calibre [ˈkælɪbə] *n.* (*Mech.*) the diameter of the inside of a tube, such as a gun-barrel

callipers [ˈkælɪpəz] *n. pl.* (*Mech.*) an instrument for measuring the exact thickness, width or diameter of objects ▶ A pair of callipers was used to measure the calibre of the pistol.

calorie (calory) (*small* **c**) [ˈkælərɪ] *n.* (*Phys.: Meas.*) a unit of heat ▶ A calorie is the amount of heat required to raise the temperature of 1 gram of water by 1°C.

Calorie (Calory) (*capital* **C**) *n.* (*Phys.: Meas.*) a unit of measurement equal to 1,000 calories ▶ The Calorie is used to measure the energy content of food. ▶ People who wish to lose weight are advised to eat food with a low Calorie content.

calorific [ˌkæləˈrɪfɪk] *a.* (*Phys.*) having to do with heating

calorific value the amount of heat produced by the combustion of a given amount of fuel

calyx [ˈkeɪlɪks] *n.* (*Bot.*) the ring of leaves surrounding the unopened bud of a flower

cam [kæm] *n.* (*Mech.*) a projection attached to a revolving shaft to give movement to another part of a machine

camber [ˈkæmbə] *n.* (*Eng.*) the tilt of a road

camcorder ['kæmˌkɔdə] *n. (Mech.; Photo.)* a video camera and sound recorder in one unit

camouflage ['kæməflɑʒ] *n. (Biol.)* a disguise of colour and shape which makes something difficult to distinguish against its background ▶ Some insects and animals have such effective camouflage that natural predators often fail to see them.

camshaft ['kæmʃɑft] *n. (Mech.)* a shaft with a number of cams, which operates the valves of an internal combustion engine

canal¹ [kə'næl] *n. (Eng.)* a waterway constructed across land to allow boats to travel on it or to bring water for irrigation, etc. ▶ The Panama Canal links the Atlantic and Pacific.

canal² *n. (Biol.)* a passage through which food, liquid or air passes in the body of a plant or animal ▶ The passage through which food passes and is digested is known as the alimentary canal.

cancer ['kænsə] *n. (Med.)* a malignant disease caused by abnormal and uncontrollable growth of cells in the body ▶ One of the most frequent causes of death is lung cancer, often caused by heavy smoking.

canker¹ ['kæŋkə] *n. (Biol.: Med.)* a disease, caused by a parasite, which affects the mouths and ears of mammals, causing sores

canker² *(Biol.)* a disease which damages the outer layer of the branches of trees, exposing the pulp beneath

cannabis ['kænəbɪs] *n. (Bot.: Med.)* a narcotic drug obtained from the leaves and flowers of the Cannabis genus of plants ▶ The possession of cannabis is illegal in many countries.

cannabis resin a sticky resin which gives the cannabis drug its strength

cantilever ['kæntɪˌlivə] *n. (Eng.)* a bracket or beam which supports a shelf or balcony, or one end of a bridge

cantilever bridge a bridge having pairs of cantilevers resting on piers, with the end meeting or connected by girders ▶ One of the earliest and longest cantilever bridges was the Forth Rail Bridge, in Scotland.

capacitance [kə'pæsɪt(ə)ns] *n. (Elec.)* the ability of a conductor system to store an electric charge, or the amount stored, measured in farads

capacitor [kə'pæsɪtə] *n. (Elec.)* a device for storing electric charge in a circuit

capacity [kə'pæsətɪ] *n. (Elec.)* the energy output of a piece of electrical equipment

capillary¹ [kə'pɪlərɪ] *n. (Biol.)* an extremely thin and narrow blood vessel, in which arterial circulation ends and venous circulation begins ▶ There is a network of capillaries in all tissues of the human body.

capillary² *a. (Biol.)* having to do with the hair

capillary tube an extremely slender glass tube used in scientific equipment ▶ The capillary tube of a thermometer usually contains mercury.

capstan ['kæpstən] *n. (Mech.)* a revolving drum or pulley, around which a cable or chain is wound mechanically in order to raise heavy objects ▶ A ship's anchor may be raised or lowered by means of a capstan.

capsule¹ ['kæpsjul] *n. (Med.)* a small gelatine or other container in which a dose of a drug is enclosed

capsule² *n. (Aer.)* the detachable section of a space vehicle which separates from the main structure and travels through space

carapace ['kærəpeɪs] *n. (Biol.)* the hard shell that protects the backs of tortoises, turtles, and similar creatures

carat¹ ['kærət] *n. (Meas.: Metall.)* a measure of the proportion of gold in an alloy, expressed in terms of twenty-fourths ▶ A twenty-two-carat gold ring contains more gold than a nine-carat gold ring.

carat² *n. (Meas.)* a measure of the weight of precious stones ▶ The international carat now weighs 0·20 grams ▶ The Cullinan diamond, found in South Africa, weighed 3,106 carats.

carbohydrate [ˌkabə(ʊ)ˈhaɪdreɪt] *n. (Chem.)* any chemical compound containing oxygen, hydrogen and carbon ▶ Carbohydrates, such as sugar and starch, are the main sources of energy in our diet.

carbon [ˈkabən] *n. (Chem.)* a chemical element contained in coal, charcoal, diamonds, petroleum, etc. ▶ There is carbon in all living matter.

carbon dating a method of estimating the age of ancient objects by measuring the amount of radioactive carbon in them ▶ Many people are now challenging the accuracy of carbon dating as a means of estimating age.

carbon dioxide a gaseous combination of one atom of carbon with two of oxygen ▶ Carbon dioxide is a natural part of the atmosphere and of exhaled breath.

carbon monoxide a poisonous gas composed of one atom of carbon to one of oxygen ▶ Carbon monoxide is a component part of car exhausts and has contributed to many deaths.

carbonate [ˈkabəneɪt] *n. (Chem.)* a salt of carbonic acid

carboniferous [ˌkabəˈnɪfərəs] *a. (Chem.: Biol.: Geol.)* containing or producing carbon or coal ▶ Carboniferous rocks were created from the decay of vegetation which grew millions of years ago.

carboy [ˈkabɔɪ] *n.* a large glass bottle, usually green or blue and protected with wickerwork ▶ Carboys are used for storing acids and other corrosive liquids.

carburettor [ˌkabəˈretə] *n. (Mech.)* a device which regulates the input of petrol to make up the mixture of petrol vapour and air in the cylinders of an internal combustion engine

carcinogen [kɑˈsɪnədʒən] *n. (Med.)* a substance that can contribute to the development of cancer ▶ Tobacco smoke is known to contain carcinogens.

carcinoma [ˌkɑsɪˈnəʊmə] *n. (Med.)* cancer, a malignant tumour

cardiac [ˈkɑdɪək] *a. (Med.)* having to do with the heart ▶ All deaths are ultimately caused by cardiac failure.

cardiac arrest sudden cessation of the heartbeat ▶ The victim suffered a severe shock, went into cardiac arrest and had to be resuscitated by the medical team.

cardinal [ˈkɑdɪn(ə)l] *a.* main, most important ▶ The cardinal points of the compass are North, South, East and West.

cardinal number a number which expresses quantity (e.g. 1, 2, 3) as opposed to sequence (e.g. 1st, 2nd, 3rd, which are ordinal numbers)

caries [ˈkeərɪz] *n. (Med.)* decay of the teeth or bones ▶ In European countries the incidence of dental caries has decreased sharply.

carnivore [ˈkanɪvɔ] *n. (Zool.)* a flesh-eating animal ▶ Lions and tigers are carnivores.

carnivorous [kɑˈnɪvərəs] *a. (Biol.)* flesh-eating ▶ Insect-eating plants are also carnivorous.

carotid [kəˈrɒtɪd] *a. (Biol.)* having to do with the carotid arteries ▶ The two carotid arteries, one on each side of the neck, supply blood to the head and brain.

carrion [ˈkærɪən] *n. (Biol.)* the decaying flesh of a dead animal ▶ Vultures do not kill, but they feed on carrion.

cartilage [ˈkɑtɪlɪdʒ] *n. (Biol.)* a strong, elastic, animal tissue, gristle ▶ Footballers often damage the cartilages in their knees.

cartography [kɑˈtɒgrəfɪ] *n. (Geog.)* the science and art of map-making

cartridge [ˈkɑtrɪdʒ] *n. (Mech.)* a cardboard and metal tube holding the exact charge to fire a bullet from a gun ▶ He had used up all his cartridges, so his rifle was of no use to him.

casing [ˈkeɪsɪŋ] *n. (Mech.)* an outer protective covering ▶ Fragments of the bomb casing were found a considerable distance from where it had fallen.

cassette [kæˈset] *n. (Mech.)* a small, usually oblong, plastic box containing a tape recording, video tape, film etc. ▶ Audio cassettes replaced vinyl records but are now themselves being displaced by compact discs.

cast [kɑst] *v.t.* (*Metall.*) to pour metal or glass into a mould, so that it sets in the shape of the mould ▶ Busts of many famous people are cast in bronze.

cast-iron [ˈkɑstˌaɪən] made of impure iron cast in a mould ▶ Many London parks used to be surrounded by ornate cast-iron railings.

catalysis [kəˈtælɪsɪs] *n.* (*Chem.*) the speeding up of a process through the action of a catalyst

catalyst [ˈkætəlɪst] *n.* (*Chem.*) an agent which speeds up a chemical process while itself remaining chemically unaffected

catalytic [ˌkætəˈlɪtɪk] *a.* (*Chem.*) concerned with a catalyst ▶ The catalytic qualities of any catalyst become weaker after prolonged use.

catalytic converter a device which can be fitted to the exhaust pipe of a motor vehicle to remove toxic gases from the exhaust fumes ▶ All new car models must now be fitted with catalytic converters.

cataract [ˈkætərækt] *n.* (*Med.*) a disease of the eyeball in which it becomes opaque, leading to loss of vision

catarrh [kəˈtɑ] *n.* (*Med.*) inflammation of the nose and throat, producing mucus

catchment [ˈkætʃmənt] *n.* a surface on which water may be caught and collected

catchment area [ˈkætʃmənt ˌeərɪə] an area of land which drains into a river or lake

caterpillar¹ [ˈkætəˌpɪlə] *n.* (*Biol.*) the larva of a certain kind of insect ▶ Caterpillars of the common cabbage-white butterfly had eaten away all the leaves of the young plants.

caterpillar² *n.* (*Mech.*) a system in which motor vehicles are fitted with continuous belts instead of wheels

caterpillar track a belt that revolves around several wheels, as on a military tank ▶ Caterpillar tracks enable heavy vehicles to move over rough or wet surfaces.

catheter [ˈkæθɪtə] *n.* (*Med.*) a tube used to drain fluid from the body, especially the bladder

cathode [ˈkæθəʊd] *n.* (*Elec.*) the negative electrode of a battery

cathode ray [ˈkæθəʊd ˌreɪ] a stream of electrons emitted from the surface of a cathode during an electrical discharge

cathode ray tube a vacuum tube in which cathode rays can be projected on to a fluorescent screen, making up a visual image ▶ The cathode ray tube is the vital element of a television set.

caustic [ˈkɔstɪk] *n.* or *a.* (*Chem.*) able to corrode or burn by chemical action, or a substance which can do this

caustic soda a strong alkaline substance (sodium hydroxide) used industrially in a number of manufacturing processes, including making soap ▶ A common domestic use of caustic soda is for unblocking drains.

cavity [ˈkævətɪ] *n.* a hole or space inside a solid object ▶ The dentist used a new amalgam to fill the cavity in one of the patient's teeth.

cavity wall a wall consisting of two layers of building material with a space between them

cc (*abbr.*) cubic centimetre

CD (*abbr.*) compact disc

cell¹ [sel] *n.* (*Biol.*) a microscopic unit of living substance ▶ The human body is made up of many thousands of millions of tiny cells.

cell² *n.* (*Elec.*) a division of an electric battery, having one pair of galvanic plates

celluloid [ˈseljʊlɔɪd] *n.* (*Chem.*) a flammable material formerly used in the manufacture of various domestic goods, toys, etc., but now largely replaced by more durable and non-combustible plastics ▶ Celluloid is still used to make films.

cellulose [ˈseljʊˌləʊs] *n.* (*Biol.*) a starchy substance that forms the cell walls of all plants ▶ Compounds of cellulose are used in the manufacture of photographic film, varnish and some textiles, as well as in the making of plastics and explosives.

Celsius [ˈselsɪəs] *see* **centigrade**

centi- *comb. form* a hundred, one-hundredth

centigrade [ˈsentɪɡreɪd] *n.* (*Phys.*) a scale for

measuring temperature ▶ On the centigrade scale, the freezing-point of water is taken as zero (0°C) and the boiling-point as 100°C.

centilitre [ˈsentɪˌliːtə] *n. (Meas.)* one-hundredth of a litre

centimetre [ˈsentɪˌmiːtə] *n. (Meas.)* one-hundredth of a metre

centipede [ˈsentɪˌpiːd] *n. (Biol.)* an insect with a large number of legs (but not 100!)

centrifugal [ˈsentrɪfjuɡ(ə)l] *a. (Phys.)* tending to pull away from the centre ▶ When an object revolves, the centrifugal force makes its component parts tend to fly away from the centre.

centrifuge [ˈsentrɪfjuːdʒ] *n. (Mech.)* a machine for separating liquids such as cream and milk by centrifugal force

centripetal [senˈtrɪpɪt(ə)l] *a. (Phys.)* tending to move towards the centre ▶ The force which draws a revolving body towards the centre is called centripetal force.

cerebellum [ˌserɪˈbeləm] *n. (Biol.)* the portion of the brain beneath the rear part of the cerebrum ▶ The cerebellum is the part of the brain that controls balance and the coordination of muscular movement.

cerebral [ˈserɪbˈb(ə)l] *a. (Med.)* having to do with the brain ▶ A common cause of sudden death is cerebral haemorrhage, or stroke.

cerebrum [ˈserɪbrəm] *n. (Biol.)* the front and principal part of the brain ▶ The cerebrum is divided into two hemispheres, which control different functions of the body.

cervical [ˈsɜːvɪk(ə)l] *a. (Biol.: Med.)* having to do with the cervix or neck

cervical smear a test designed to detect early signs of cancer of the cervix ▶ Women are encouraged to have cervical smears at regular intervals.

cervical spine the vertebrae in the neck

cervix [ˈsɜːvɪks] *n. (Biol.: Med.)* the neck of the womb

chain [tʃeɪn] *n. (Phys.: Chem.)* a series of atoms linked together in a molecule

chain reaction a self-perpetuating chemical or nuclear reaction ▶ In a chain reaction, each step is triggered by the preceding one and itself causes the next in the series, which is therefore difficult to stop.

chamber [ˈtʃeɪmbə] *n. (Mech.)* an enclosed or confined space in a machine or mechanism ▶ There was only one bullet left in the chamber of his revolver.

charge¹ [tʃɑːdʒ] *v.i. and v.t. (Elec.)* to accumulate electricity or to cause electricity to be accumulated in something ▶ The batteries of my cordless telephone are charged overnight, when it is not in use.

charge² *n. (Elec.)* the amount or accumulation of electricity in a battery, etc.

chassis [ˈʃæsɪ] *n. (Mech.)* the base or framework of a vehicle or appliance to which the other parts are attached ▶ The body of the car was badly dented in the collision but the chassis remained undamaged.

chemical [ˈkemɪk(ə)l] *n. and a. (Chem.)* having to do with or produced by chemistry, or a substance produced by chemical means ▶ The chemical industry employs thousands of people. ▶ Indiscriminate use of chemicals on the land can be dangerous to health.

chemical change the formation of a new substance by chemical reaction

chemical reaction the process of changing one substance into another by chemical means

chemical symbol a letter or letters used to signify a chemical element ▶ H is the chemical symbol for hydrogen, and O for oxygen. ▶ Both chemical symbols are used in the formula for water – H_2O.

chemistry [ˈkemɪstrɪ] *n.* the scientific study of substances and how they react to each other

chemotherapy [ˌkiːməʊˈθerəpɪ] *n. (Chem.: Med.)* the treatment of diseases, especially cancer, by the use of chemicals, i.e. drugs

chickenpox [ˈtʃɪkɪnˌpɒks] *n. (Med.)* a contagious disease, usually occurring during childhood ▶ Chickenpox is characterized by a high temperature and a rash of itchy spots on the skin.

chill [tʃɪl] *n. (Med.)* a cold, shivering sensation caused by a drop in body temperature

chill factor [ˈtʃɪl ˌfæktə] a weather effect in which the temperature is felt to be lower than it actually is because of the influence of other factors, especially the wind (*also called* **wind chill factor**)

chiropody [kɪˈrɒpədɪ] *n.* (*Med.*) the science of treatment of the feet

chiropractor [ˈkaɪrəʊˌpræktə] *n.* (*Med.*) an expert in spinal manipulation as a cure for pain or disease ▶ Chiropractors operate in the field of alternative medicine.

chloral [ˈklɔr(ə)l] *n.* (*Chem.*) a narcotic liquid made from chlorine and alcohol ▶ Chloral is sometimes used as an anaesthetic.

chloride [ˈklɔraɪd] *n.* (*Chem.*) a compound of chlorine and another element

chlorinate [ˈklɔrɪˌneɪt] *v.t.* (*Chem.*) to add chlorine to something, e.g. the water in a swimming pool, as a disinfectant ▶ The smell and taste of chlorinated water are typical of municipal swimming pools.

chlorine [ˈklɔrin] *n.* (*Chem.*) a yellowish-green gas with a strong smell, obtained from common salt and used to disinfect water

chloroform [ˈklɒrəfɔm] *n.* (*Chem.*: *Med.*) a volatile liquid formerly used as an anaesthetic

chlorophyll [ˈklɒrəfɪl] *n.* (*Biol.*: *Chem.*) the green substance in plants that absorbs sunlight and facilitates growth

choke [tʃəʊk] *n.* (*Mech.*) a device for controlling the flow of air into a petrol engine ▶ New cars now have automatic chokes to replace the old manual variety. ▶ Starting a cold engine requires use of the choke.

cholera [ˈkɒlərə] *n.* (*Med.*) an acute bacterial infection spread by contaminated water ▶ Cholera causes severe vomiting and diarrhoea, leading to dehydration.

cholesterol [kəˈlestər(ɒ)l] *n.* (*Biol.*: *Med.*) a fatty substance found in blood and body organs, and thought, in excess, to be the cause of hardening of the arteries ▶ Sufferers from heart disease are advised to cut down their intake of cholesterol.

chord [kɔd] *n.* (*Maths.*) a straight line joining the ends of an arc or two points on a curve

chromatics [krəˈmætɪks] *n. sg.* (*Phys.*) the science of colour

chrome *see* **chromium**

chromium [ˈkrəʊmɪəm] *n.* (*Chem.*: *Metall.*) a hard, shiny metallic element often used as an alloy with other metals ▶ Many metal surfaces on bicycles and other manufactured goods are plated with chromium to prevent rust.

chromosome [ˈkrəʊməˌsəʊm] *n.* (*Biol.*) one of the rod-shaped structures in living cells that carry the genes that transmit hereditary characteristics ▶ Some congenital diseases result from chromosome abnormalities.

chronic [ˈkrɒnɪk] *a.* (*Med.*) (of a disease) long-lasting ▶ She has suffered for many years from chronic rheumatism.

chrysalis [ˈkrɪsəlɪs] *n.* (*Biol.*) a moth or butterfly in an early stage of its development, when it is enclosed in a hard, protective covering, or this protective covering itself ▶ When a butterfly emerges from its chrysalis, its wings are wet and have to become dry before it can fly.

circuit [ˈsɜkɪt] *n.* (*Elec.*) the track along which an electric current passes from one terminal to another ▶ The lights failed because a faulty connection had broken the circuit.

circulate [ˈsɜkjʊˌleɪt] *v.i.* (*Biol.*) (of the blood) to move around via the veins to the heart and via the arteries from the heart to the rest of the body

circulation [ˌsɜkjʊˈleɪʃ(ə)n] *n.* (*Biol.*: *Med.*) the movement of the blood through the body ▶ The patient developed gangrene in his toe because of poor circulation of the blood.

circum- *comb. form* around

circumference [sɜˈkʌmfər(ə)ns] *n.* (*Maths.*) a line which forms the boundary of a circle

circumscribe [ˈsɜkəmˌskraɪb] *v.t.* (*Maths.*) to draw a line around a geometrical shape in such a way that it touches all the corners

cirrhosis [sɪˈrəʊsɪs] *n.* (*Med.*) the development of fibrous tissue in the liver, leading to loss of function and eventual failure ▶ Excessive consumption of alcohol may lead to cirrhosis.

cistern [ˈsɪstən] n. (Mech.) a water tank

citrus [ˈsɪtrəs] n. (Biol.) any of the genus of trees that includes oranges, lemons, etc.

civil [ˈsɪvl] a. having to do with commercial or administrative affairs

civil engineering the branch of engineering concerned with building public works, such as roads, railways. bridges, canals and harbours

cladding [ˈklædɪŋ] n. (Mech.: Build.) a protective coating of stone (on a building) or of insulating material (on a pipe or tank)

clamp¹ [klæmp] n. (Mech.) a piece of apparatus which can be adjusted, usually by a screw, to hold several items firmly together ▶ The wooden boards were held firmly in a clamp until the adhesive had set.

clamp² v.t. (Mech.) to fix a clamp on the wheel of an illegally parked car so that it cannot move off

class [klɑs] n. (Biol.) the second highest division in the taxonomy of plants and animals ▶ Class comes below phylum and above order in the biological classification.

claustrophobia [ˌklɔstrəˈfəʊbɪə] n. (Psych.) morbid fear of being in a confined space ▶ She was afraid to enter the caves because of her tendency to claustrophobia.

clearance [ˈklɪər(ə)ns] n. (Eng.) the space between a moving object and another object ▶ The engineers had to leave enough clearance beneath the bridge for trucks to pass under it.

cline [klaɪn] n. a continuous gradation of differences, e.g. temperatures

clinic [ˈklɪnɪk] n. (Med.) a department of a hospital or other establishment where medical advice and treatment, often of a specialized nature, are available ▶ Because he was suffering from continual headaches, he made an appointment at the eye clinic.

clinical [ˈklɪnɪk(ə)l] a. (Med.) (of medicine) concerned directly with patients

clone [kləʊn] n. (Biol.) one of a number of organisms which have been reproduced asexually from a plant or animal ▶ A clone is an exact copy of the organism from which it has been reproduced.

closed [kləʊzd] a. (Elec.) (of an electrical circuit) unbroken, so that the current can flow

closed circuit television a television system for a restricted number of viewers, often in one building, in which the picture is transmitted by cable ▶ The bank robbery was clearly visible to the security guards on the closed circuit television.

clot [klɒt] n. (Biol.: Med.) (of the blood) a soft mass of coagulated blood ▶ A blood clot in his coronary artery impaired the circulation and the patient suffered a heart attack.

clutch [klʌtʃ] n. (Mech.) a device for connecting and disconnecting two revolving shafts in an engine ▶ The driver released the clutch and the car moved smoothly away.

coagulant [kəʊˈægjʊl(ə)nt] n. (Chem.) a substance which helps liquid to coagulate

coagulate [kəʊˈægjʊleɪt] v.i. (Biol.) to change from a liquid into a soft mass or clot ▶ The blood on the wound soon coagulated and the bleeding stopped.

coalesce [kəʊəˈles] v.i. (Phys.) to come together and merge into one substance or mass ▶ Drops from a tap will coalesce into a pool of water.

coaxial [kəʊˈæksɪəl] a. (Maths.) having a common axis

coaxial cable a cable with two concentric conductors separated by an insulator ▶ By using coaxial cables, we can transmit a number of different signals down the same line at the same time.

cocaine [kəˈkeɪn] n. (Chem.: Med.) a drug derived from the leaves of the coca plant and used as a stimulant or as a local anaesthetic

cocoon [kəˈkun] n. (Biol.) a silky covering spun by the larvae of various insects ▶ The larvae of butterflies and moths remain in their cocoons until they are ready to fly.

code [kəʊd] *n. (Comput.)* a system of symbols by which data is loaded into a memory and which can be read and decoded for access

codeine [ˈkəʊdiːn] *n. (Biol.: Med.)* a drug obtained from opium and used as a narcotic and a pain-killer

coefficient[1] [ˌkəʊɪˈfɪʃ(ə)nt] *n. (Maths.)* a multiplying number ▶ In the expression 2xy, the coefficient of xy is 2 and the coefficient of y is 2x.

coefficient[2] *n. (Phys.)* a measure of a property of a substance, e.g. expansion or viscosity, in particular circumstances

cog [kɒg] *n. (Mech.)* one of the teeth around the edge of a cog-wheel in a machine ▶ The cogs fit into similar cogs on another wheel, so that the one cannot turn without the other.

cog-wheel [ˈkɒɡ ˌwiːl] a wheel with teeth, or cogs, around the rim

cohesion [kəʊˈhiːʒ(ə)n] *n. (Phys.)* the force which holds the molecules of a substance together

coil [kɔɪl] *n. (Elec.)* a wire wound around a central core to provide resistance or inductance

col [kɒl] *n. (Meteor.)* an area of low pressure between two anticyclones

cold-blooded [ˌkəʊldˈblʌdɪd] having a blood temperature that changes according to the temperature of the environment ▶ Fish and snakes are cold-blooded.

cold front the front edge of an advancing mass of cold air ▶ A cold front moved slowly across the country, bringing dry but chilly weather.

colic [ˈkɒlɪk] *n. (Med.)* acute pain in the bowels, stomach-ache

colitis [kɒˈlaɪtɪs] *n. (Med.)* inflammation of the colon

colon [ˈkəʊlən] *n. (Biol.: Med.)* the lower part of the large intestine

coma [ˈkəʊmə] *n. (Med.)* a long period of absolute unconsciousness ▶ After the accident he was in a coma for almost a week.

comatose [ˈkəʊmətəʊs] *a. (Med.)* in a state of coma

combustible [kəmˈbʌstəbl] *n. and a. (Chem.: Phys.)* able and liable to burn

combustion [kəmˈbʌstʃ(ə)n] *n. (Chem.)* the combination of a substance with oxygen or another element, producing light and heat; burning

comet [ˈkɒmɪt] *n. (Astron.)* a heavenly body with a shining head and a long tail which follows an eccentric orbit around the Sun ▶ One of the most famous bodies of this sort is Halley's comet, which can be seen clearly when it passes close to Earth.

community [kəˈmjuːnɪtɪ] *n. (Ecol.)* a set of interdependent plants and animals inhabiting a given area ▶ Disturbance or destruction of any one element in an ecological community will have an effect also on the remainder of its inhabitants.

compact disc (CD) [ˌkɒmpæk(t) ˈdɪsk] a small audio disc which is read by laser and on which a great amount of digitally-encoded data is stored on a small surface ▶ Compact video discs are also available.

companionway [kəmˈpænjənˌweɪ] *n. (Naut.)* a stairway from one deck to another on a ship

compass [ˈkʌmpəs] *n. (Phys.: Mech.)* an instrument, indicating magnetic north and south, used for finding direction, especially on ships and aircraft

compasses [ˈkʌmpəsɪz] *n. pl. (Maths.: Geog.)* an instrument for drawing circles or measuring distances between two points, e.g. on a map ▶ A pair of compasses is an essential tool for a navigator.

compatible [kəmˈpætəbl] *a. (Med.: Comput.)* capable of being used together ▶ The two men's blood groups were not compatible, so a transfusion could not be given. ▶ The two computers were perfectly compatible, so he was able to use our program.

complement [ˈkɒmplɪmənt] *n. (Maths.)* (of an angle) one which when added to a given angle makes up ninety degrees (90°) ▶ An angle of 40 degrees (40°) is the complement to an angle of 50 degrees (50°).

complementary [ˌkɒmplɪˈment(ə)rɪ] *a. (Maths.)*

(of an angle) being the complement of another ▶ An angle of 25 degrees (25°) is complementary to one of 65 degrees (65°) – (25 + 65 = 90).

complex[1] [kɒmpleks] *n.* (*Psych.*) an abnormal condition of the mind resulting from repressed desire or disagreeable experiences in the past ▶ After a series of unlucky experiences in his youth, he developed a persecution complex, and thought that everyone was against him.

complex[2] *n.* (*Build.*) a number of buildings associated with one another and serving the same purpose in the same place ▶ A shopping complex was built on the old school playing fields.

component [kəm'pəʊnənt] *n.* (*Mech.*) one of the parts of which something is made up ▶ When the apparatus was assembled, it was discovered that a vital component was missing.

composite ['kɒmpəzɪt] *n.* (*Bot.*) a member of a family of plants whose heads are composed of a number of small flowers

composite number a product of two other numbers greater than 1, the opposite of a prime number

compost ['kɒmpɒst] *n.* (*Biol.: Hort.*) a mixture of decaying vegetable matter and chemicals, for use as fertilizer ▶ Many gardeners keep compost heaps in a corner of the garden.

compound ['kɒmpaʊnd] *n.* (*Chem.*) a combination of two or more elements formed by chemical action ▶ Water (H_2O) is a compound of two atoms of hydrogen with one of oxygen.

compound fracture [ˌkɒmpaʊnd 'fræktʃə] a fracture in which the skin surrounding the break in the bone is damaged, often with the bone protruding

compress [kəm'pres] *v.t.* (*Phys.: Mech.*) to press together into a smaller space ▶ In a car engine, the piston compresses a mixture of petrol and air, which explodes when ignited. ▶ The brakes of heavy vehicles are sometimes operated by compressed air.

compression [kəm'preʃ(ə)n] *n.* (*Phys.: Mech.*) the process or result of compressing, or a measure of the pressure resulting from the process ▶ The compression chamber was leaking and the engine was losing power.

compressor [kəm'presə] *n.* (*Mech.*) a device for compressing air or another gas ▶ A centrifugal compressor is used in some types of jet engine.

comptometer [kɒmp'tɒmɪtə] *n.* (*Maths.: Mech.*) a kind of calculating machine

compute [kəm'pjut] *v.t.* (*Maths.: Comput.*) to calculate, including by use of a computer

computer [kəm'pjutə] *n.* (*Maths.: Mech.*) an electronic device which performs mathematical calculations or processes data loaded into it at great speed according to the instructions contained in a program

computer game a game requiring great speed of reaction and coordination of hand and eye, in which the player reacts via a keyboard or joystick to graphics on a screen

computer hacker someone who gains illegal or unauthorized access to a computer system

computer language the code or language used in a computer program

computer program a set of instructions fed into a computer, usually via a disk

computer science the science of constructing and using computers

computer system one of a number of versions of computer hardware and the compatible programs for use with it

computer virus *see* **virus**[2]

computer graphics visual images produced on a computer screen and capable of being manipulated via a keyboard

computerize[1] [kəm'pjutəraɪz] *v.t.* (*Maths.: Mech.: Comput.*) to instal computers to replace other means of controlling data ▶ The central banks are now fully computerized.

computerize[2] *v.t.* (*Comput.*) to load a corpus of data into a computer system ▶ The records of all previous business have now been computerized.

con [kɒn] *v.t.* (*Naut.*) to direct the steering of a ship

concave [ˈkɒŋkeɪv] *a. (Maths.)* having a surface curving inwards, like the inside of an arc of a circle

concentrate¹ [ˈkɒnsəntreɪt] *v.t. (Chem.)* to increase the strength of a substance by reducing its volume, e.g. by removing water ▶ The concentrated acid was safely stored in carboys.

concentrate² *n. (Chem.)* a substance in concentrated form

concentric [kənˈsentrɪk] *a. (Maths.)* (of circles) having the same centre ▶ The radio waves were depicted as moving outward from the transmitter in a series of concentric circles.

concrete [ˈkɒŋkrit] *n. (Build.)* a mixture of cement, gravel, sand and water ▶ Concrete may be poured into moulds while it is wet, but is extremely hard when it dries.

concussion [kənˈkʌʃ(ə)n] *n. (Med.)* temporary injury to the brain as caused by a blow on the head ▶ After the collision the driver was taken to the hospital suffering from severe concussion.

condensation [ˌkɒndenˈseɪʃ(ə)n] *n. (Phys.)* the process or result of the reduction of a substance into a more compact form, e.g. when gas or vapour turns into liquid ▶ It was difficult to see through the windows because of the condensation of water vapour from the boiler on the cold glass.

condense¹ [kənˈdens] *v.t. (Phys.)* to change the form of a substance from a gas to a liquid

condense² *v.t. (Phys.)* to make a liquid stronger and more viscous by reducing its volume

condenser¹ [kənˈdensə] *n. (Phys.)* an apparatus for reducing steam to a liquid form

condenser² *n. (Elec.)* a device for acccumulating electricity

conditioned reflex a natural response to a stimulus which, because of repetitive training, becomes also the reponse to a different stimulus

condom [ˈkɒndɒm] *n. (Med.)* a contraceptive sheath worn by men during sexual intercourse ▶ As well as having a contraceptive function, the condom provides protection against the transmission of venereal disease.

conduct [kənˈdʌkt] *v.t. (Phys.: Elec.)* (of a substance) to allow electricity to flow along it or to pass on heat ▶ Metals conduct electricity, but wood does not.

conduction [kənˈdʌkʃ(ə)n] *n. (Phys.: Elec.)* the passing along of heat or electricity

conductivity [ˌkɒndʌkˈtɪvɪtɪ] *n. (Phys.: Elec.)* a measure of the ability of a substance to allow heat or electricity to pass through it ▶ The conductivity of copper is very high.

conductor [kənˈdʌktə] *n. (Phys.: Elec.)* a substance which allows heat or electricity to pass through it ▶ Stone is a poor conductor of electricity.

conduit [ˈkɒndɪt] *n. (Elec.: Mech.)* a tube or pipe through which electric wires or cables pass ▶ The conduits for the electric wiring are concealed beneath the floorboards.

cone [kəʊn] *n. (Maths.)* a solid figure with a round, flat base and a pointed top ▶ For some reason, long stretches of the motorways are often cordoned off with plastic cones.

configuration [kənˌfɪgjʊˈreɪʃ(ə)n] *n. (Astron.)* the relative positions of the planets at any given time

confluence [ˈkɒnfluəns] *n. (Geog.)* the joining of two rivers ▶ Baghdad is situated at the confluence of the Tigris and the Euphrates.

congeal [kənˈdʒil] *v.i. (Phys.)* (of a fluid) to change to a solid because of the action of cold ▶ It was so cold that even the lubricating oil had congealed in the tank.

congenital [kənˈdʒenɪt(ə)l] *a. (Biol.: Med.)* existing in the genes from before birth and thus inherited from the parents ▶ By genetic engineering, medical researchers may in the future be able to eliminate certain congenital diseases.

congested [kənˈdʒestɪd] *a. (Med.)* abnormally full of blood or another fluid ▶ His breathing was impaired because his lungs were congested.

conglomerate [kənˈglɒmərət] *n. (Geol.)* rock consisting of small stones held together by clay

congruent ['kɒŋgruənt] *a. (Maths.)* (of geometrical figures) having the same shape ▶ The diagram consisted of a series of congruent triangles.

conical ['kɒnɪk(ə)l] *a. (Maths.)* cone-shaped

conjunction [kən'dʒʌŋkʃ(ə)n] *n. (Astron.)* the Earth and another planet are said to be in conjunction when the Sun and the planet are in a straight line as seen from the Earth

conjunctiva [ˌkɒndʒʌŋ(k)'taɪvə] *n. (Biol.: Med.)* the mucous membrane that lines the inside of the eyelids and the front of the eyeball

conjunctivitis [kənˌdʒʌŋ(k)tɪ'vaɪtɪs] *n. (Med.)* inflammation of the conjunctiva

connect [kə'nekt] *v.t. (Mech.)* to join two things together, as when a household supply is joined to the main water or electricity supply ▶ The old cottage was not yet connected to the mains and relied on the well in the garden for its water.

connecting rod [kə'nektɪŋ ˌrɒd] a rod that transmits power from one part of a machine to another ▶ In a car engine, power is transmitted from the piston to the crankshaft by a connecting rod.

connection [kə'nekʃ(ə)n] *n. (Mech.)* an object or device which joins two sections of an apparatus together ▶ Water was leaking from the pipe because of a faulty connection.

conning-tower ['kɒnɪŋˌtaʊə] the superstructure of a submarine, from which the steering and firing of weaponry are directed ▶ At first only the periscope was visible, but then the conning tower rose above the surface of the sea.

conservancy [kən'sɜvənsɪ] *n. (Ecol.)* official preservation of forests, fisheries etc.

conservation [ˌkɒnsə'veɪʃ(ə)n] *n. (Ecol.)* protection of cultural monuments and of natural resources and the environment from human destruction ▶ A vigorous campaign is under way to ensure the conservation of the rainforests of the Amazon basin.

conserve [kən'sɜv] *v.t. (Ecol.)* to protect from despoliation

consistency [kən'sɪstənsɪ] *n. (Chem.)* the degree of cohesion or density of a substance

console ['kɒnsəʊl] *n. (Elec.: Mech.)* the control panel of an electrical or an electronic system ▶ The producer monitored the recording from the studio console.

constant ['kɒnstənt] *n. or a. (Maths.)* a value that remains unchanged ▶ The equation contained a number of variables, but the value of *x* remained constant.

constellation [ˌkɒnstə'leɪʃ(ə)n] *n. (Astron.)* a number of fixed stars grouped within the outline of an imaginary figure in the sky ▶ Among the best known constellations are the Great Bear – which is also called the Plough – and the Southern Cross.

contact ['kɒntækt] *n. (Elec.)* the touching of two points in an electric circuit, allowing the current to flow

contact breaker ['kɒntækt ˌbreɪkə] a device that breaks an electrical circuit at regular intervals, creating a spark that ignites the mixture in a petrol engine

contagious [kən'teɪdʒəs] *a. (Med.)* (of an illness) spread by touch ▶ Leprosy is such a highly contagious disease that sufferers used to be kept in separate colonies away from the rest of the population.

contaminate [kən'tæmɪneɪt] *v.t. (Phys.: Chem.: Med.)* to pollute by the addition of foreign matter ▶ The crops were contaminated with radioactive fallout. ▶ Many beaches have been contaminated by sewage.

contamination [kənˌtæmɪ'neɪʃ(ə)n] *n. (Chem.: Med.)* pollution

contour ['kɒntʊə] *n. (Geog.)* a line on a map joining places of equal height above sea-level

contraception [ˌkɒntrə'sepʃ(ə)n] *n. (Med.)* birth control

contraceptive [ˌkɒntrə'septɪv] *n. or a. (Med.)* a device used in birth control

control [kən'trəʊl] *n. (Mech.)* a device which governs the operation of a machine, such as a pedal or a gear-lever ▶ The co-pilot took over the controls when the captain of the aircraft became unwell.

control tower [kənˈtrəʊl ˌtaʊə] the airport building from which air traffic is controlled

convalesce [ˌkɒnvəˈles] v.i. (Med.) to recover health after an operation or a severe illness

convalescence [ˌkɒnvəˈles(ə)ns] n. (Med.) a period in which people rest and recover

convalescent [ˌkɒnvəˈles(ə)nt] a. or n. (Med.) a person who is convalescing

convection [kənˈvekʃ(ə)n] n. (Phys.) the transmission of heat or electricity through liquids and gases by the movement of heated particles

convector [kənˈvektə] n. (Mech.) a heater which works by circulating currents of hot air

convert [kənˈvɜt] v.i. and v.t. (Maths.: Mech.) to change from one system, form or use to another ▶ To convert from kilometres to miles you divide by eight and multiply by five. ▶ The central heating system has been converted from oil to gas.

convex [ˈkɒnveks] a. (Maths.) curving outward, like the outside of an arc of a circle

convexity [kənˈveksɪtɪ] n. (Maths.) the state of being convex

conveyor [kənˈveɪə] n. (Mech.) a device for transporting things from one place to another

conveyor belt [kənˈveɪə ˌbelt] an endless mechanical belt which conveys components along an assembly line in a manufacturing process

convulsion [kənˈvʌlʃ(ə)n] n. (Med.) a spasmodic movement of the muscles ▶ People suffering from epilepsy sometimes have convulsions.

coolant [ˈkuːlənt] n. (Mech.) a liquid used to keep a machine with moving parts cool ▶ Water is used as a coolant in car engines.

coordinate [kəʊˈɔdɪnət] n. (Maths.: Geog.) a system of lines, often with numbers or letters, used to pinpoint an exact position on a map or chart

copper [ˈkɒpə] n. or a. (Chem.: Metall) a red, malleable metallic element notable as an excellent conductor of electricity ▶ The circuit was closed with a section of insulated copper wire.

core[1] [kɔ] n. (Geol.) the central part of the Earth, beneath the crust

core[2] n. (Eng.) the round mass of earth and rock brought up by a hollow drill, e.g. when drilling for oil

core[3] n. (Elec.) the central part of a cable carrying electric wires

core[4] n. (Elec.) a piece of soft iron or other magnetic material inside an induction coil

core[5] n. (Phys.) the part of a nuclear reactor in which the fissile material is contained

cornea [ˈkɔnɪə] n. (Biol.: Med.) the transparent outer covering of the front of the eye

corona [kəˈrəʊnə] n. (Astron.) the outermost part of the Sun's atmosphere ▶ Except during an eclipse of the sun, the corona cannot be seen.

coronary [ˈkɒrən(ə)rɪ] a. (Med.) resembling a crown ▶ The term coronary is frequently used as shorthand for coronary thrombosis in the sense of heart attack.

coronary arteries the two main arteries which supply blood to the heart

coronary thrombosis the formation of a blood clot in one of the arteries of the heart

corpus [ˈkɔpəs] n. (Comput.) a body or mass ▶ The researchers amassed a huge corpus of data to be loaded into the computer for analysis.

corpuscle [ˈkɔpʌsl] n. (Biol.: Med.) a red or white blood cell ▶ White corpuscles play an important part in the defence of the body against harmful bacteria.

corrode [kəˈrəʊd] v.i. and v.t. (Chem.) to wear away through chemical action ▶ The acid spilt from the car battery had corroded the metal leads and broken the electrical circuit.

corrosion [kəˈrəʊʒ(ə)n] n. (Chem.) the process or result of corroding

corrosive [kəˈrəʊsɪv] a. (Chem.) referring to a substance which corrodes another

corrugated [ˈkɒrəgeɪtɪd] a. (Metall.) (of iron) pressed into waves or folds and galvanized ▶ Many small buildings and sheds are roofed with corrugated iron.

cortex [ˈkɔteks] *n. (Biol.)* the outer layer of an organ, especially the brain or the kidney ▶ The cortex of the brain is the cerebral cortex, and that of the kidney the renal cortex.

cortisone [ˈkɔtɪzəʊn] *n. (Chem.: Med.)* a drug used to reduce inflammation ▶ Cortisone is used in the treatment of rheumatoid arthritis.

cosmic [ˈkɒzmɪk] *a. (Astron.: Phys.)* concerned with the cosmos

cosmology [kɒzˈmɒlədʒɪ] *n. (Astron.: Phys.)* the study of the nature and origin of the cosmos

cosmonaut [ˈkɒzmənɔt] *n. (Aer.)* an astronaut, someone who travels in a spaceship beyond the Earth's atmosphere ▶ The world's first cosmonaut was the Russian, Yuri Gagarin.

cosmos [ˈkɒzmɒs] *n. (Astron.: Phys.)* the universe

couple [ˈkʌpl] *v.t. (Mech.)* to join two things together ▶ The sleeping car was coupled on to the train at the next station.

coupling [ˈkʌplɪŋ] *n. (Mech.)* a device for joining two things together ▶ The coupling snapped and the sleeping car remained at the platform.

course¹ [kɔs] *n. (Med.)* a programme of treatment prescribed by a doctor to alleviate or cure a disease ▶ The patient was given a course of antibiotics to cure the infection.

course² *n. (Build.)* one layer of bricks etc. in a wall

crane [kreɪn] *n. (Eng.)* a machine for lifting and lowering heavy weights ▶ The long jibs of the dockside cranes were visible for miles.

cranium [ˈkreɪnɪəm] *n. (Biol.)* the part of the skull that contains the brain

crank [kræŋk] *n. (Mech.)* part of an axle or shaft bent at right angles in order to convert movement up and down into circular movement, or vice versa ▶ Old cars had to be started by turning a crank handle.

creosote [ˈkrɪəsəʊt] *n. (Chem.)* a liquid distilled from coal-tar and used to preserve wood ▶ The new fence was painted with creosote to protect it from the weather.

cretaceous [krɪˈteɪʃəs] *a. (Geol.)* like or containing chalk ▶ The layers of cretaceous rocks are thought to date from the Cretaceous period, some 150 million years ago.

cross [krɒs] *n. (Biol.)* an animal or plant engendered by two different species ▶ A mule is a cross between a horse and a donkey.

cross-breed [ˈkrɒsˌbrid] to mate animals of two different kinds; the progeny of such mating ▶ The mule is a cross-breed of a horse and a donkey.

cross-fertilization [ˌkrɒsfɜtɪlaɪˈzeɪʃ(ə)n] the production of flowers by applying the pollen of one kind to the pistil of another kind

cross-section [ˈkrɒsˌsekʃ(ə)n] a slice cut across a stem or organ for purposes of analysis ▶ A cross-section of the diseased stem was examined through a microscope to identify the nature of the disease.

crowbar [ˈkrəʊbɑ] *n. (Mech.)* a metal (usually iron) bar with one hooked end ▶ The lids of the crates were prised off with a heavy crowbar.

crucible [ˈkrusɪbl] *n. (Chem.: Metall.)* a heat-resistant container in which metals and other substances may be heated to very high temperatures ▶ Small crucibles are used in chemistry laboratories, but the basin at the bottom of a furnace in which molten metals are collected is also called a crucible.

crude [krud] *a. (Chem.: Metall.)* (of raw materials) in a natural, untreated state

crude oil oil as it emerges from a well, before it is refined

crust [krʌst] *n. (Geol.)* the hard, solid, outer part of the Earth

crustacean [krʌsˈteɪʃ(ə)n] *n. (Biol.: Zool.)* an animal, such as a crab or a lobster, which has jointed limbs and a body partly covered by a shell ▶ Most crustaceans live in water.

cryogenics [ˌkraɪəʊˈdʒenɪks] *n. sg. (Phys.)* the physics of substances at very low temperatures

cryometer [kraɪˈɒmɪtə] *n. (Phys.)* a thermometer for measuring temperatures near or below freezing

crystal [ˈkrɪst(ə)l] *n. (Chem.)* a small, regularly shaped piece of a mineral, usually colourless and transparent

crystalline [ˈkrɪstəlaɪn] *a. (Chem.)* made of or referring to crystals

crystallize [ˈkrɪstəlaɪz] *v.i.* and *v.t. (Chem.)* to form into crystals

crystallographer [ˌkrɪstəˈlɒɡrəfə] *n. (Chem.)* someone who studies the nature and shapes of crystals

cube¹ [kjub] *n. (Maths.)* a solid object with six square and equal planes

cube² *n. (Maths.)* a number multiplied by itself twice ▶ The cube of 3 is 27 ($3 \times 3 \times 3$).

cube root the number which has been multiplied by itself twice in order to produce the cube ▶ The cube root of 125 is 5 (125/25/5).

cubic¹ [ˈkjubɪk] *a. (Maths.)* cube-shaped

cubic² *a. (Maths.)* having the volume of a cube with sides of a given length ▶ A cubic centimetre of a substance contains the volume equal to a cube of which each side measures one centimetre.

cull [kʌl] *v.t. (Zool.)* to restrict the number of animals (e.g. in a herd) by killing a selected number ▶ Many people now oppose the practice of culling as cruel.

cultivate [ˈkʌltɪveɪt] *v.t. (Agric.)* to raise plants or animals in a garden or farm

culture [ˈkʌltʃə] *n. (Biol.: Med.)* controlled growth of bacteria or fungi in a laboratory for experimental or therapeutic use ▶ Cultures of certain highly dangerous bacteria have to be kept under conditions of strict security.

cuneate [ˈkjuniət] *a. (Maths.)* wedge-shaped

current¹ [ˈkʌrənt] *n. (Elec.)* the flow of an electric charge along a conductor ▶ The operator switched on the current and the engine began to hum.

current² *n. (Phys.)* a movement of water or air faster or in a different direction from the water or air around it ▶ The rapid current in the narrow river swept the little boat along. ▶ A sudden upward current of air lifted the parachute over the top of the tree.

cut out [ˌkʌtˈaʊt] (of an engine) to cease functioning because of a fault ▶ Suddenly, the engine cut out and the car rolled to a halt.

cut out [ˈkʌt ˌaʊt] *n. (Elec.: Mech.)* a device for breaking an electric circuit when it becomes overloaded ▶ When too many appliances are switched on at the same time, the cut-out comes into operation to prevent overheating and the risk of fire.

cutting¹ [ˈkʌtɪŋ] *n. (Eng.)* a wide passage cut through high ground to allow the passage of a road, railway, canal etc. ▶ It is said that for every ten metres of the railway cutting, one worker was accidentally killed.

cutting² *n. (Hort.)* a section of a growing plant used in propagation ▶ Rows of cuttings in pots lined the shelves of the greenhouse.

cwt [ˈhʌndrədˌweɪt] *(abbr.)* hundredweight, a twentieth part of a ton

cyanide [ˈsaɪənaɪd] *n. (Chem.)* a highly poisonous chemical compound

cybernetics [ˌsaɪbəˈnetɪks] *n. sg. (Phys.: Biol.: Mech.)* the comparative study of control and communication mechanisms in machines and living creatures

cycle [ˈsaɪkl] *n. (Elec.)* one complete series of changes in a quantity that varies at regular intervals, as in an alternating electric current

cyclometer [saɪˈklɒmɪtə] *n. (Mech.)* an instrument for measuring the distance travelled by a bicycle etc., by counting the number of revolutions of a wheel of known circumference

cylinder [ˈsɪlɪndə] *n. (Mech.)* a hollow object with round, flat ends and straight sides ▶ In a car engine the mixture is injected into cylinders for compression and ignition.

cylindrical [sɪˈlɪndrɪkl] *a. (Maths.)* shaped like a cylinder

cyst [sɪst] *n. (Med.)* a growth or sac in the body, containing morbid matter ▶ She was admitted to hospital for the removal of an ovarian cyst.

d

daisy-wheel [ˈdeɪzɪˌwil] a wheel-shaped computer printer with the characters on spikes around the perimeter

damp [dæmp] *a.* moist, humid

damp course [ˈdæmp ˌkɔs] a layer of impervious material laid between courses near the bottom of a wall to prevent damp from rising up it

damper [ˈdæmpə] *n. (Mech.)* a movable metal plate which can be adjusted to control the flow of air to a fire

dashboard [ˈdæʃbɔd] *n. (Mech.)* the instrument panel of a car

dashlight [ˈdæʃlaɪt] *n. (Mech.)* the light illuminating the dashboard of a car

data [ˈdeɪtə] *n. sg.* or *pl. (Comput.)* facts from which information can be inferred, as in a computer program ▶ The data was loaded into the mainframe computer and accessible at all stations in the network.

data bank *see* **database**

database [ˈdeɪtə ˌbeɪs] *n. (Comput.)* a mass of data fed into a computer and easily accessible ▶ The database at the hospital contained the medical records of all the patients.

data processing [ˌdeɪtə ˈprəʊsesɪŋ] the loading and processing of data in computer files

davit [ˈdævɪt] *n. (Naut.)* one of a pair of beams used as cranes for lowering lifeboats from the side of a ship

dB *(abbr.)* decibel

DC *(abbr.)* direct current

de- *comb. form* reversal or separation

decaffeinate [diˈkæfɪneɪt] *v.t.* to remove the caffeine from coffee ▶ People who tend to sleep badly often prefer to drink only decaffeinated coffee.

decibel [ˈdesɪbel] *n. (Phys.)* a unit expressing the comparative loudness of sounds ▶ The noise of the aircraft taking off was measured in decibels and declared to be excessive.

decimal [ˈdesɪm(ə)l] *a. (Maths.)* having to do with tens or tenths ▶ The decimal system of counting now in use is in most countries of the world.

decimal currency a monetary system in which coins or notes represent the given unit in multiples of ten ▶ There are 10 cents in a dime and 10 dimes (100 cents) in a dollar.

decimal fraction a fraction (e.g. $\frac{1}{2}$) expressed in tenths (0·5)

decimal point [ˈdesɪm(əl ˌpɔɪnt] the point or comma on the left of the decimal, as in 3·5 or 3,5

decimalization [ˌdesɪməlaɪˈzeɪʃ(ə)n] *n. (Maths.)* the process or result of being converted to the decimal system ▶ First proposals for the decimalization of the British currency were strongly opposed.

decimalize [ˈdesɪməlaɪz] *v.t. (Maths.)* to convert to the decimal system ▶ British currency was decimalized in 1971.

decode [diˈkəʊd] *v.i.* and *v.t. (Comput.)* to read and transfer a coded message into plain language

decompose [ˌdikəmˈpəʊz] *v.i.* and *v.t. (Biol.: Phys.)* (of light, a substance etc.) to break up or separate into its parts ▶ A dead body decomposes very quickly, especially in hot weather. ▶ A prism decomposes light into its constituent colours. ▶ Unstable compounds may decompose if they get wet.

decompress [ˌdikəmˈpres] *v.t. (Phys.: Mech.)* to reduce the pressure of air or a gas to normal atmospheric pressure

decompression [ˌdikəmˈpreʃ(ə)n] *n. (Phys.: Mech.)* the process or result of decompressing

decompression chamber [dikəm'preʃ(ə)n ˌtʃeɪmbə] a chamber in which a person (e.g. a deep-sea diver) is gradually returned to normal atmospheric pressure ▶ Failure to make proper use of the decompression chamber can cause severe pains and problems with breathing.

decontaminate [ˌdikən'tæmɪneɪt] *v.t.* (*Chem.: Med.*) to clear of poisonous substances or radioactivity

defecate ['defəkeɪt] *v.i.* (*Biol.*) to expel faeces from the body

deflate [di'fleɪt] *v.i.* and *v.t.* (*Phys.*) (of a tyre, balloon, etc.) to let down or go down by allowing air or gas to escape ▶ The balloon was punctured by the branch of a tree and as the helium escaped it rapidly deflated. ▶ The rubber tube was deflated and stowed away.

defoliant [ˌdi'fəʊlɪənt] *n.* (*Chem.*) a chemical agent used to kill leaves ▶ An orange defoliant was sprayed on the trees from the air.

defoliate [ˌdi'fəʊlɪeɪt] *v.t.* (*Chem.*) to strip of leaves ▶ Whole areas of tropical forest were defoliated by chemical means during the fighting in the jungle.

degradable [dɪ'greɪdəbl] *a.* (*Biol.: Chem.*) capable of being decomposed biologically or chemically ▶ Environmentalists encourage the use of degradable materials for packaging, etc.

degrade [dɪ'greɪd] *v.t.* (*Biol.: Chem.: Geol.: Phys.*) to reduce to a lower grade ▶ Rocks are degraded by weather into pebbles and sand. ▶ Molecules can be degraded into atoms.

degree¹ [dɪ'gri] *n.* (*Phys.: Meas.*) a unit of temperature ▶ Water boils at 100 degrees centigrade (Celsius) (100°C).

degree² *n.* (*Maths.: Meas.*) a measure of the size of angles ▶ 90 degrees (90°) make up a right angle.

degree³ *n.* (*Maths.: Geog.: Meas.*) a measure of latitude and longitude ▶ The ship was located at a point roughly 10 degrees west (10°W) of its scheduled route.

dehydrate [di'haɪdreɪt] *v.t.* (*Chem.*) to remove water from something ▶ Dehydrated foodstuffs, such as milk and eggs, can be stored for long periods.

dehydration [ˌdihaɪ'dreɪʃ(ə)n] *n.* (*Chem.: Med.*) the process or result of loss of water ▶ The cholera patient suffered severe dehydration and was given intravenous fluids.

deliquesce [ˌdelɪ'kwes] *v.i.* (*Chem.*) to liquefy by absorbing moisture from the atmosphere

delirious [dɪ'lɪərɪəs] *a.* (*Psych.*) suffering from delirium ▶ At the height of the fever he became delirious and it was impossible to understand what he said. ▶ She was delirious with joy.

delirium [dɪ'lɪərɪəm] *n.* (*Psych.*) wandering of the mind, often with delusions and loss of touch with reality

delirium tremens [dɪˌlɪərɪəm 'trimenz] an acute stage of alcoholism

delivery [dɪ'lɪvərɪ] *n.* (*Biol.*) the process of birth of a human being or other mammal ▶ The father asked to be present at the delivery of the baby.

delta ['deltə] *n.* (*Geog.*) an area of land, shaped more or less like a triangle, where a river divides into a number of branches before flowing into the sea

delta rays electrons moving at relatively low speeds

dendrochronology [ˌdendrə(ʊ)krə'nɒlədʒɪ] *n.* (*Biol.*) the study of the annual growth rings of trees ▶ Dendrochronology can be used to date historical events.

denominator [dɪ'nɒmɪneɪtə] *n.* (*Maths.*) the part of a fraction below the line ▶ The denominator in $\frac{3}{4}$ is 4.

density ['densətɪ] *n.* (*Phys.*) the amount of matter in relation to the volume of a substance ▶ Metals have a greater density than wood.

dentistry ['dentɪstrɪ] *n.* (*Med.*) the science of the treatment and care of the teeth

deodorant [di'əʊdərənt] *n.* (*Chem.*) a substance which counteracts unpleasant smells, especially that of perspiration ▶ Failure to use a deodorant is considered antisocial.

deoxidize [diˈɒksɪdaɪz] *v.i.* and *v.t.* *(Chem.)* to lose or remove oxygen

deposit [dɪˈpɒzɪt] *n.* *(Geol.)* matter accumulated by natural means ▶ Geologists believe that there are vast deposits of mineral ores on the moon.

depression¹ [dɪˈpreʃ(ə)n] *n.* *(Meteor.)* an area of low pressure, signifying bad weather ▶ A deep depression was centred over the British Isles for most of the summer.

depression² *n.* *(Geog.)* a hollow or low-lying area of land ▶ Rain water collected in the depressions in the ground.

depression³ *n.* *(Psych.)* lowering of mood ▶ She dissolved in floods of weeping due to depression brought on by the awful weather.

depressive [dɪˈpresɪv] *a.* and *n.* *(Psych.)* causing or characterized by depression, or a person suffering from depression ▶ The behaviour of a manic depressive is characterized by periods of unreasonable happiness alternating with fits of depression.

dermatitis [ˌdɜməˈtaɪtɪs] *n.* *(Med.)* inflammation of the skin

derrick [ˈderɪk] *n.* *(Eng.)* a simple kind of crane, used especially in ports

derv [dɜv] *n.* *(Chem.)* diesel engine fuel oil ▶ The word derv consists of the first letters of diesel engine road vehicle.

desalinate [diˈsælɪneɪt] *v.t.* *(Chem.)* to remove salt from sea-water ▶ Many new industrial plants are needed to desalinate the water and irrigate the desert.

descale [diˈskeɪl] *v.t.* *(Chem.)* to remove by a chemical reaction the deposits of chalk, etc., left by water with a high content of lime ▶ The kettle has to be descaled regularly because of the hard water.

desensitize [diˈsensətaɪz] *v.t.* *(Chem.: Med.)* to render insensitive to something ▶ She was allergic to cats and was given a course of vaccine to desensitize her.

desiccate [ˈdesɪkeɪt] *v.t.* and *v.i.* *(Chem.)* to dry up

desiccator [ˈdesɪkeɪtə] *n.* *(Chem.)* a piece of apparatus for drying substances that tend to decompose if allowed to get wet

detergent [dɪˈtɜdʒ(ə)nt] *n.* *(Chem.)* a chemical agent for cleaning or washing clothes, etc.

detonate [ˈdetəneɪt] *v.t.* *(Chem.: Mech.)* to set off an explosion

detonator [ˈdetəneɪtə] *n.* *(Mech.)* a device for detonating an explosive charge

detritus [dɪˈtraɪtəs] *n.* *(Geol.)* loose material, such as sand and stones, resulting from the wearing away of rock

deviant [ˈdiviənt] *a.* *(Psych.: Biol.)* (of behaviour, nature, etc.) different from the norm

deviate [ˈdivieɪt] *v.i.* *(Psych.: Biol.)* to differ from the norm ▶ His behaviour was unpredictable and occasionally deviated widely from what was socially acceptable.

device [dɪˈvaɪs] *n.* *(Mech.)* a mechanism designed to enable a specific task to be done

deuterium [djuˈtɪərɪəm] *n.* *(Phys.: Chem.)* a heavy kind of hydrogen, used in hydrogen bombs

dia- *comb. form* through, apart, across

diabetes [ˌdaɪəˈbitiz] *n.* *(Med.)* a disease of the pancreas caused by insulin deficiency

diabetic [ˌdaɪəˈbetɪk] *a.* and *n.* *(Med.)* having to do with or suffering from diabetes ▶ The standard treatment of diabetics is by injections of insulin.

diagnose [ˌdaɪəgˈnəʊz] *v.t.* *(Med.)* to determine the nature and cause of an illness ▶ The patient complained of chest pains and the doctor diagnosed pneumonia.

diagnosis [ˌdaɪəgˈnəʊsɪs] *n.* *(Med.)* a determination of the nature and cause of a disease

diagonal [daɪˈægən(ə)l] *n.* *(Maths.)* a straight line from one corner of a figure to the opposite corner ▶ The diagonal of a rectangle runs through the centre.

dialysis [daɪˈælɪsɪs] *n.* *(Med.)* the filtering of blood to remove waste products ▶ People suffering from kidney failure may be given dialysis via a kidney machine.

diameter [daɪˈæmɪtə] *n.* *(Maths.)* a straight line through the centre of a circle ▶ The diameter of the planet Jupiter is nearly twelve times larger than the diameter of Earth.

diaphragm¹ [ˈdaɪəfræm] *n. (Elec.)* the vibrating disc in the mouthpiece and earpiece of a telephone or in the loudspeaker of a radio

diaphragm² *n. (Med.)* a contraceptive device used by women

diaphragm³ *n. (Med.)* the layer of muscle between the chest and the abdomen

diarrhoea [ˌdaɪəˈrɪə] *n. (Med.)* frequent discharge of liquid faeces from the bowels ▶ He was suffering from diarrhoea and vomiting caused by food poisoning.

diesel [ˈdizl] *n. (Eng.)* a vehicle fuelled by diesel oil (*see* **derv**)

diesel engine a type of internal combustion engine in which explosions are caused by spraying oil into compressed and heated air ▶ Diesel engines are commonly used in lorries, buses, and other heavy-duty vehicles.

diet [daɪət] *n. (Med.)* a prescribed course of food and drink followed for health reasons ▶ He was advised to adhere to a strict diet in order to lose weight.

dietetics [ˌdaɪəˈtetɪks] *n. sg. (Med.)* the scientific study of diet and nutrition

dietician [ˌdaɪəˈtɪʃ(ə)n] *n. (Med.)* a specialist in dietetics

differential [ˌdɪfəˈrenʃ(ə)l] *a. (Maths.: Mech.)* based on a system of differences

differential calculus *(Maths.)* the method of finding an infinitely small quantity which, when multiplied by itself an infinite number of times, will equal a given quantity

differential gear a gear which enables the back wheels of a (back-wheel drive) car to revolve at different speeds when rounding a corner

digit [ˈdɪdʒɪt] *n. (Maths.)* the written symbol for any of the numbers 0–9 ▶ The date 1066 contains four digits

digital [ˈdɪdʒɪt(ə)l] *a. (Maths.: Mech)* using numbers ▶ On a digital watch, the time is given in a series of numbers (e.g. 12:45) instead of in conventional symbols around the clock face.

dilate [daɪˈleɪt] *v.t. (Med.)* to enlarge a bodily cavity for purposes of examination ▶ The drug dilated the pupils of his eyes so that the optician could examine them more easily.

dilute [daɪˈljut] *v.t. (Chem.)* to make a liquid thinner by adding water ▶ The concentrated sulphuric acid was diluted for use in the experiment.

dimension [daɪˈmenʃ(ə)n] *n. (Meas.)* an aspect of something which can be measured ▶ It is important to know the height, width, volume and other such dimensions before you start work.

DIN [ˌdiaɪˈen] *(abbr.) (Photo.)* a measure of the speed of photographic film by its sensitivity to light ▶ The higher the DIN number, the faster the film. ▶ The term DIN is an abbreviation of the German *Deutsche Industrie Norm* (German Industrial Norm.).

diphtheria [dɪfˈθɪərɪə] *n. (Med.)* a serious infection of the throat

dipsomania [ˌdɪpsəʊˈmeɪnɪə] *n. (Psych.)* an irresistible craving for alcohol ▶ Dipsomania is now more commonly called alcoholism.

dipstick [ˈdɪpˌstɪk] *n. (Mech.)* a calibrated metal rod, dipped into a container to check the level of liquid in it ▶ The dipstick showed that the level of oil in the engine was low.

direct [daɪˈrekt] *a. (Maths.: Elec.)* proceeding from one point to another in a straight line

direct current an electric current that flows in only one direction

direction finder [dəˈrekʃ(ə)n ˌfaɪndə] a device for determining the direction from which radio signals are being transmitted

directional aerial an aerial adjusted to receive radio signals from a specific direction

discharge¹ [dɪsˈtʃɑdʒ] *n. (Elec.)* loss of an electrical charge ▶ Accidental contact with a metal surface led to a dangerous discharge of current from the battery.

discharge² *n. (Med.)* pus or other infected matter leaking from a wound

discharge³ *v.t. (Eng.)* to emit waste matter ▶ The power station was feared to have discharged radioactive waste into the sea.

disease [dɪˈziz] *n. (Med.)* illness, sickness, an unhealthy condition resulting from infection, etc. ▶ Coughs and sneezes spread diseases.

disinfect [ˌdɪsɪnˈfekt] *v.t. (Chem.: Med.)* to clean from infection, often by chemical means ▶ The huts were disinfected before they could be reoccupied after the epidemic was over.

disinfectant [ˌdɪsɪnˈfektənt] *n. (Chem.: Med.)* a chemical substance used to disinfect things

dislocate [ˈdɪsləkeɪt] *v.t. (Med.)* (of a bone) to put out of joint ▶ The rugby player dislocated his shoulder when he was tackled.

dispensary [dɪsˈpens(ə)rɪ] *n. (Med.)* a place where medicines are prepared and distributed

displacement [dɪsˈpleɪsmənt] *n. (Phys.)* the water displaced by a floating body ▶ In measuring displacement, the weight of water displaced equals the weight of the floating body at rest.

dissect [dɪˈsekt] *v.t. (Biol.)* to cut up a plant or an animal's body in order to examine its structure ▶ They dissected a frog in the laboratory and sketched internal organs.

distil [dɪˈstɪl] *v.t. (Chem.)* to purify a liquid by heating it until it becomes a vapour, removing any impurities deposited, and then cooling it slowly until it becomes a liquid again ▶ Only distilled water, free from impurities, should be used in a car battery.

distributor [dɪˈstrɪbjʊtə] *n. (Elec.: Mech.)* the device in a petrol engine that distributes electric current to the sparking plugs

divide [dɪˈvaɪd] *v.t. (Maths.)* to find out how many times one number is contained in (will go into) another ▶ 2 goes into 10 five times ($10 \div 2 = 5$).

diving bell [ˈdaɪvɪŋ ˌbel] a hollow vessel, formerly bell-shaped, in which people are lowered into deep water to observe marine life, having air piped from the surface

division [dɪˈvɪʒn] *n. (Maths.)* the process of dividing one number by another

DNA [ˈdienˈeɪ] *(abbr.) (Biol.: Chem.)* deoxyribonucleic acid, the main constituent of chromosomes, which transmits inherited characteristics ▶ The discovery of DNA has been described as unlocking the secret of life.

dorsal [ˈdɔs(ə)l] *a. (Zool.)* situated on the back, e.g. of a fish ▶ The dorsal fin helps the fish to maintain its balance.

dose [dəʊs] *n. (Med.)* the amount of a medicine prescribed for use at one time ▶ Take one dose, three times daily.

drag [dræg] *n. (Aer.)* the resistance to movement experienced by an aircraft in flight ▶ The thrust of its engines drives an aircraft forward, and the wings provide the lift, but both thrust and lift have to overcome drag.

drain¹ [dreɪn] *v.t.* to take away superfluous fluid ▶ The blood was drained from the incision by a suction pipe, so that the surgeon could have a better view.

drain² *v.t. or n. (Eng.)* to take away fluid by a channel, or the channel itself ▶ Surface rain-water runs off through drains into the sewers.

draw [drɔ] *v.t. (Metall.)* to stretch or lengthen ▶ Soft metals, such as copper, are suitable to be drawn into wires for carrying electric current.

drive [draɪv] *n. (Mech.)* the means of operating a machine, such as a car ▶ Cars made for use in Britain have right-hand drive because they travel on the left side of the road.

drug [drʌg] *n. (Chem.: Med.)* a chemical substance used in the treatment of illnesses ▶ One of the most common drugs in everyday use is aspirin ▶ Drugs may be harmful if improperly used.

drug addict [ˈdrʌg ˌædɪkt] a person addicted to narcotic drugs

drug pusher [ˈdrʌg ˌpʊʃə] a person who sells narcotic drugs illegally

drum¹ [drʌm] *n. (Mech.)* a cylinder which may be used as a container or around which a hawser or wire may be wound ▶ Oil is stored in large metal drums ▶ The winch cable was wound around a wooden drum.

drum² *n. (Med.)* the part of the inner ear that vibrates when struck by sound waves ▶ The noise of the explosion was so loud that it almost shattered the drums of my ears.

dry [draɪ] *a.* without moisture

dry battery an electric battery made up of dry cells

dry cell a battery cell in which the electrolyte is a paste and not a liquid

dry dock a dock from which the water can be drained for the repair of ships

dub¹ [dʌb] *v.t. (Elec.)* to add a new soundtrack to a film ▶ The French film was dubbed into English.

dub² *v.t. (Elec.)* to copy a recording from one tape to another ▶ The album was illegally dubbed and sold in the street-market.

dubbing master [ˈdʌbɪŋ ˌmɑstə] a high-quality recording, from which multiple copies are dubbed

duct [dʌkt] *n. (Mech.: Biol.)* a tube through which a gas or a liquid is conveyed ▶ Warm air was circulated via a series of ducts beneath the floor.

ductile [ˈdʌktaɪl] *a. (Metall.)* (of a metal etc.) which can be easily drawn out into thread ▶ Copper is extremely ductile and therefore suitable for use in making wire.

duodenal [ˌdjuə(ʊ)ˈdin(ə)l] *a. (Med.)* having to do with the duodenum ▶ Abdominal pains are frequently caused by duodenal ulcers.

duodenum [ˌdjuə(ʊ)ˈdinəm] *n. (Biol.)* the first part of the small intestine

dynamics [daɪˈnæmɪks] *n. sg. (Phys.)* the scientific study of force, movement, and energy

dynamo [ˈdaɪnəməʊ] *n. (Elec.: Mech.)* a machine for turning mechanical energy into electrical energy by a process of induction

dysentery [ˈdɪs(ə)ntrɪ] *n. (Med.)* an infection of the bowels causing diarrhoea and dehydration

dyslexia [ˌdɪsˈleksɪə] *n. (Psych.)* a neurological disorder affecting ability to read and write

e

earth [ɜθ] *v.t.* (*Elec.*) to use the Earth as part of an electrical circuit by running a wire from the plug down into the ground ▶ As a safety measure, all the electrical circuits should be properly earthed.

earth science a branch of science having the Earth as the subject of study, e.g. Geography and Geology

ebb [eb] *v.i.* to fall, decrease ▶ After a week without rain, the floodwaters began to ebb.

ebb tide the tide as it is going out

eccentric¹ [ɪk'sentrɪk] *a.* or *n.* unusual, not like others

eccentric² *n.* (*Mech.*) a mechanical device for converting circular motion into movement backward and forward, as on a steam engine

eccentric circles circles having different centres, i.e. not concentric

eccentric orbit an orbit, followed by a planet or satellite, which is not circular

eccentric rod a rod transmitting the motion of an eccentric wheel

eccentric wheel a wheel that does not turn about its centre

ECG (*abbr.*) electrocardiogram, electrocardiograph

echo ['ekəʊ] *n.* (*Phys.*) a reflection of light or sound off a solid object

echo chamber ['ekəʊˌtʃeɪmbə] a room with walls that echo sounds for recording or radio effects

echo sounder ['ekəʊ ˌsaʊndə] an apparatus for sounding the depth of water beneath the keel of a ship

eclipse [ɪ'klɪps] *n.* (*Astron.*) the total or partial blocking of light from a heavenly body by the passage of another body between it and the eye of the viewer ▶ An eclipse may be partial, when only a part of the body is obscured, or total.

ecological [ˌi:kə'lɒdʒɪk(ə)l] *a.* having to do with ecology

ecological chain the interdependence of the inhabitants of an ecological community ▶ The use of insecticides protects crops from the damage caused by insects, but it disrupts the ecological chain by robbing insectivorous birds of their natural source of food.

ecologist [i'kɒlədʒɪst] *n.* a specialist in ecology

ecology [i'kɒlədʒɪ] *n.* the relationship between living things and their environment, or the study of this ▶ Environmentalists are concerned at the effect on the ecology of the oceans of the dumping of nuclear waste.

economics [ˌikə'nɒmɪks] *n. sg.* the study of the production, distribution, and consumption of goods and of the commercial activities of society

ecosystem ['ikəʊˌsɪstəm] *n.* (*Ecol.*) a system consisting of a community of organisms and its environment ▶ The ecosystem of the area was disrupted by the destruction of the forest.

effervesce [ˌefə'ves] *v.i.* (*Chem.*) to bubble up or to escape in bubbles ▶ Alcoholic liquids effervesce during the process of fermentation.

effervescence [ˌefə'ves(ə)ns] *n.* (*Chem.*) the state or process of effervescing

effervescent [ˌefə'ves(ə)nt] *a.* (*Chem.*) having to do with effervescence

effluent ['efluənt] *n.* (*Chem.*) the liquid that is discharged from a sewage tank ▶ The release of effluent into rivers and lakes may damage or destroy fish stocks.

effusion [ɪ'fjuːʒn] *n.* a sudden and unrestrained

pouring out of liquid or gas ▶ The explosion resulted in effusions of water and gas from ruptured pipes in the building.

ejector seat [ɪˈdʒektə ˌsit] a seat that can be shot clear of an aeroplane in case of emergency ▶ Many pilots owe their lives to the use of ejector seats when their aircraft went out of control.

elastic [ɪˈlæstɪk] *a.* (*Chem.*) made of material that will return to its former shape after being stretched or otherwise distorted

elasticity [ˌɪlæsˈtɪsətɪ] *n.* (*Chem.*) the quality of being elastic ▶ The elasticity of the materials used made them less liable to break under stress.

electric [ɪˈlektrɪk] *a.* containing, generating, or operated by electricity

electric cable an insulated wire for conveying an electric current

electric charge the accumulation of electric energy in a battery

electric circuit the passage of electricity by means of a conductor

electric field a region in which forces are exerted on any electric charge present

electric shock sudden pain felt when electricity passes through the body

electricity [ˌɪlekˈtrɪsətɪ] *n.* (*Phys.*) a powerful agent that produces heat and light, as well as a number of other effects on different materials ▶ The discovery and use of electricity has enabled us to heat and light our homes without open fires or flames.

electrification [ɪˌlektrɪfɪˈkeɪʃ(ə)n] *n.* (*Phys.*) conversion of a mechanical system to work by electricity ▶ The electrification of the railway marked the end of the age of steam.

electrify [ɪˈlektrɪfaɪ] *v.t.* (*Phys.*) to charge with electricity

electro- *comb. form* driven by or relating to electricity

electrocardiogram [ɪˈlektrəʊˈkɑdɪə(ʊ)ˈkɑdɪəgræm] *n.* (*Med.*) a record produced on an electrocardiograph

electrocardiograph [ɪˌlektrə(ʊ)ˈkɑdɪəgræf] *n.* (*Med.*) an instrument which shows and records heartbeats

electrocute [ɪˈlektrəkjut] *v.t.* (*Elec.*) to kill by electric shock ▶ He accidentally touched a high-tension cable and was electrocuted.

electrode [ɪˈlektrəʊd] *n.* (*Phys.*) one of the poles of an electric battery

electrolysis [ˌɪlekˈtrɒləsɪs] *n.* (*Phys.*) the decomposition of chemical compounds by having an electric current passed through them

electrolyte [ɪˈlektrə(ʊ)ˌlaɪt] *n.* (*Chem.*: *Phys.*) a chemical compound that can be decomposed by having an electric current passed through it

electromagnet [ɪˌlektrə(ʊ)ˈmægnɪt] *n.* (*Phys.*) a bar of soft iron made magnetic by having an electric current passed through it

electromagnetic radiation [ɪˌlektrə(ʊ)mægˌnetɪk reɪdɪˌeɪˈʃ(ə)n] radiation consisting of an electric and a magnetic field, at right angles to each other, and including radio waves, gamma rays, X-rays, short-wave light etc.

electromotive force mechanical motion produced by electricity ▶ Rechargeable batteries provide the electromotive force for the new, environment-friendly buses.

electron [ɪˈlektrən] *n.* (*Phys.*) a particle bearing a negative electric charge

electronic [ˌɪlekˈtrɒnɪk] *a.* (*Phys.*) having to do with, or produced by, electronics

electronics [ˌɪlekˈtrɒnɪks] *n. sg.* (*Phys.*) the science of applied physics that deals with the conduction of electricity in a vacuum or a semiconductor

electrostatic [ɪˌlektrə(ʊ)ˈstætɪk] *a.* (*Elec.*) having to do with, producing or produced by static electricity

element[1] [ˈelɪmənt] *n.* (*Chem.*) one of the fundamental substances of which something is composed, which cannot be further reduced by chemical analysis ▶ The table of chemical elements shows substances which cannot be broken down into other substances.

element[2] *n.* (*Elec.*) the resistance wire in an electric appliance, creating heat ▶ One of the elements in the electric fire had burned out, reducing the heat emitted from it.

elementary [elɪˈment(ə)rɪ] *a.* simple, uncomplicated, basic

elementary particle [elɪˌment(ə)rɪ ˈpɑtɪk(ə)l] any one of such entities as electrons, neutrons or protons, which are less complex than atoms and are considered incapable of subdivision

elevation¹ [ˌelɪˈveɪʃ(ə)n] *n.* *(Geog.)* height above sea level

elevation² *n.* *(Astron.)* the height of a heavenly body above the horizon, stated as an angle

elevator¹ [ˈelɪˌveɪtə] *n.* *(Mech.)* a machine for raising large quantities of small objects, such as grain, from one level to another ▶ The grain was transferred by elevators from the ship's hold into lorries.

elevator² *n.* *(Aer.)* a movable control surface on the tail of an aeroplane which is raised to make it climb and lowered to make it dive

ellipse [ɪˈlɪps] *n.* *(Maths.)* a regular oval shape

elliptical [ɪˈlɪptɪkl] *a.* *(Maths.)* in the shape of an oval, egg-shaped ▶ The satellite went into an elliptical orbit around the earth.

embryo [ˌembrɪəʊ] *n.* *(Biol.)* a young plant or animal in the early stage of development before birth from an egg or its mother's uterus

embryology [ˌembrɪˈɒlədʒɪ] *n.* *(Biol.)* the scientific study of embryos

embryonic [ˌembrɪˈɒnɪk] *a.* *(Biol.)* referring to an embryo in a very early stage of development

emetic [ɪˈmetɪk] *n.* *(Med.)* a preparation that causes vomiting ▶ The child who had eaten poisonous berries was given an emetic to empty its stomach.

EMF [ˌiemˈef] *(abbr.)* electromotive force

emission [ɪˈmɪʃ(ə)n] *n.* the process or result of emitting or being emitted

emit [ɪˈmɪt] *v.t.* *(Elec.: Mech.: Eng.)* to give off, send out ▶ The loudspeaker emitted a series of squawks and shrieks. ▶ A beam of intensely brilliant light was emitted from the laser. ▶ A cloud of noxious fumes was emitted from the factory chimneys.

emulsion¹ [ɪˈmʌlʃ(ə)n] *n.* *(Chem.)* a liquid suspended in another liquid so that they will not completely separate ▶ Emulsion paint contains a resin and water.

emulsion² *n.* *(Chem.)* a light-sensitive substance used for coating films ▶ When the film was accidentally exposed, the emulsion absorbed the light and made it unusable.

endemic [enˈdemɪk] *a.* *(Biol.: Med.)* peculiar to a particular locality or people ▶ Certain diseases are endemic in some areas but rarely spread to others.

endocrine gland a ductless gland which pours a specific secretion into the bloodstream

endocrinology [ˌendə(ʊ)krɪˈnɒlədʒɪ] *n.* *(Biol.: Med.)* the study of the secretions of endocrine glands

energy [ˈenədʒɪ] *n.* *(Phys.)* the power of a body to perform mechanical work ▶ In the modern industrial world the main source of energy is electricity.

engine [ˈendʒɪn] *n.* *(Mech.)* an apparatus with a number of moving parts, which supplies the power to perform various kinds of work ▶ Engines range from simple pulleys to the most powerful aero-engines.

ENT [ˌienˈti] *(abbr.)* Ear, Nose and Throat (a branch of medicine or department of a hospital)

entomology [ˌentəˈmɒlədʒɪ] *n.* *(Biol.)* the study of insects

environment [ɪnˈvaɪ(ə)rənmənt] *n.* *(Biol.)* the sum of external influences affecting any living organism ▶ The pollutants emitted by industrial plants pose a severe threat to the environment.

environmental [ɪnˌvaɪ(ə)rənˈment(ə)l] *a.* having to do with the environment ▶ Environmental factors are increasingly being taken into account when new building projects are being planned.

environmentalist [ɪnˌvaɪ(ə)rənˈment(ə)lɪst] *n.* a person who is concerned with the study and preservation of the environment ▶ The efforts of environmentalists have managed to save many natural habitats of plants and animals from destruction by road-builders.

enzyme ['enzaɪm] *n. (Biol.)* a substance produced by living cells which promotes the working of internal organs, especially the digestive system

epi- *comb. form* at, to, besides, in addition

epicentre ['epɪsentə] *n. (Geol.)* the point on the surface of the Earth directly above the place where an earthquake has its origin ▶ The epicentre was far out to sea but the tidal wave caused by the earthquake flooded many coastal towns.

epidemic [epɪ'demɪk] *n. (Med.)* an outbreak of a disease that attacks many people at the same time and spreads with great rapidity ▶ The influenza epidemic killed almost as many people as the war.

epilepsy ['epɪlepsɪ] *n. (Med.)* a disorder of the brain that causes convulsions of the body and sometimes loss of consciousness

epileptic [ˌepɪ'leptɪk] *a. (Med.)* referring to epilepsy ▶ As a girl she suffered from occasional epileptic fits, which decreased as she grew older.

equation [ɪ'kweɪʒ(ə)n] *n. (Maths.)* two algebraic expressions equal to one another, as shown by the symbol = ▶ If $x = 2y + 3$, and $y = 1\frac{1}{2}$, then $x = 6$.

equator [ɪ'kweɪtə] *n. (Geog.)* an imaginary line drawn around the surface of the Earth, equidistant at all points from both poles ▶ The equator runs through the middle of Africa.

equi- *comb. form* equal

equilateral [ˌikwɪ'læt(ə)r(ə)l] *a. (Maths.)* having all sides of equal length ▶ The most obviously equilateral figure is a square.

equilibrium [ˌikwɪ'lɪbrɪəm] *n. (Mech.)* a state of rest or balance resulting from the fact that all forces acting on something are equal and opposite

equinox ['ikwɪnɒks] *n. (Astron.)* the moment when the sun crosses the equator so that night and day are of equal length

equivalent [ɪ'kwɪvələnt] *a. (Chem.: Geol.: Maths.)* having the same combining power, position, or area, as something else

erg [ɜg] *n. (Phys.)* a unit of measurement of work done

erode [ɪ'rəʊd] *v.t. (Geol.)* to wear or rub away ▶ Over several thousands of years the wind and rain eroded the soil down to the bare rock.

erosion [ɪ'rəʊʒ(ə)n] *n. (Geol.)* wearing or rubbing away ▶ The erosion of the soil because of unscientific farming has turned large areas of the world's surface into desert.

ersatz [ɜ'zæts] *a.* imitation, artificial ▶ Ersatz coffee made from acorns was very bitter.

erupt [ɪ'rʌpt] *v.i. (Geol.)* (of a volcano) to throw up lava, steam etc. ▶ When Mount Vesuvius erupted in AD 79, the whole of Pompeii was buried in hot ash.

eruption [ɪ'rʌpʃ(ə)n] *n. (Geol.)* (of a volcano) the throwing up of lava etc. ▶ The sudden eruption of Krakatoa caused widespread panic and destruction.

escalator ['eskəleɪtə] *n. (Mech.)* a moving staircase ▶ Travellers on the Underground are saved by the escalators from walking up and down hundreds of stairs.

ether ['iθə] *n. (Chem.: Med.)* a light, volatile and flammable fluid, produced by the distillation of alcohol with an acid and used as an anaesthetic ▶ Ether has now been generally replaced in surgery by other forms of anaesthetics.

ethnic ['eθnɪk] *a.* belonging to a specific tribe or race ▶ In a multicultural society the needs of ethnic minorities have always to be borne in mind.

ethnology [eθ'nɒlədʒɪ] *n.* the science which studies the different varieties of the human race

eugenics [ju'dʒenɪks] *n. sg. (Biol.)* the study of improving offspring by selective breeding

euthanasia [juθə'neɪzjə] *n. (Med.)* assisting the terminally ill to die painlessly ▶ Though seemingly an act of mercy, euthanasia raises great moral questions and is therefore illegal.

evaporate [ɪ'væpəreɪt] *v.i.* and *v.t. (Phys.)* to change or cause to change into steam ▶ In the heat of the midday sun, the rain puddles quickly evaporated and disappeared.

evolution [ˌivəˈluʃ(ə)n] *n.* *(Biol.)* the theory that all forms of life have developed from simpler forms or from a single rudimentary form, or the process of development put forward by this theory

excavator [ˈekskəˌveɪtə] *n.* *(Eng.)* a machine for digging deep holes in the ground ▶ The excavator digging the foundations for a disco hall discovered the remains of a Roman temple.

excision [ekˈsɪʒ(ə)n] *n.* *(Med.)* removal by cutting out, as in a surgical operation

excrete [ɪkˈskrit] *v.t.* *(Biol.)* to discharge waste matter from the body

exhale [eksˈheɪl] *v.t.* and *v.i.* *(Biol.)* to breathe out

exhaust [ɪgˈzɔst] *n.* *(Eng.)* a component of a machine enabling steam, gas, fumes etc. to escape ▶ Emissions from car exhausts contribute to the pollution of the environment.

expand [ɪkˈspænd] *v.t.* and *v.i.* *(Phys.)* to stretch, grow larger ▶ Metals expand when hot.

expansion [ɪkˈspænʃ(ə)n] *n.* *(Phys.)* the process or result of expanding ▶ The expansion of the metal moving parts in the engine led to increased friction and eventual overheating.

exponent [ɪkˈspəʊnənt] *n.* *(Maths.)* a number or quantity written as a superscript after another number to indicate how many times it is to be multiplied by itself ▶ In the expression 4^3, the exponent ^{3} shows that 4 is to be multiplied by itself 3 times ($4 \times 4 \times 4$). (*see* **power**)

express [ɪkˈspres] *v.t.* *(Maths.)* to represent a quantity by using symbols or formulae ▶ The relationship between the two quantities can be expressed as $x + 2 = y - 3$ ($y = x + 5$).

expression [ɪkˈspreʃ(ə)n] *n.* *(Maths.)* a combination of symbols representing an algebraic quantity

f

F *(abbr.)* Fahrenheit, farad
facet [ˈfæsɪt] *n.* *(Chem.)* one of the faces of a crystal or a cut gem
facsimile [fækˈsɪməlɪ] *n.* an exact copy
factor [ˈfæktə] *n.* *(Maths.)* one of the quantities that multiplied together make a product ▶ 3 and 4 are factors of 12.
faeces [ˈfiːsiz] *n. pl.* *(Biol.: Med.)* waste matter excreted from the bowels ▶ Allowing one's dog to deposit faeces in a public place is a legal offence.
Fahrenheit [ˈfɑrənhaɪt] *a.* *(Phys.)* pertaining to the temperature scale on which the freezing-point of water is marked at 32° and the boiling-point at 212° ▶ This scale is named after its inventor, Gabriel Daniel Fahrenheit.
fall-out [ˈfɔːlaʊt] the deposit of radioactive dust after a nuclear explosion ▶ Because of the strong winds, the effect of fall-out from the nuclear bomb was felt many miles away.
family [ˈfæməlɪ] *n.* *(Biol.)* a division or subdivision of a taxonomic order of plants or animals
fan [fæn] *n.* *(Mech.)* a device with radiating blades that is rotated to create a draught ▶ The car engine contains a fan to prevent it from overheating.
fan belt [ˈfæn ˌbelt] a belt which drives the radiator fan and generator in a car ▶ If a car engine overheats it is probable that the fanbelt has broken.
farad [ˈfærəd] *n.* *(Elec.)* a unit of measurement of the capacity of a condenser ▶ The name farad is derived from the name of Michael Faraday, who first defined it.
fascia [ˈfeɪʃə] *n.* *(Mech.)* the instrument panel of a car
fat [fæt] *n.* *(Biol.)* an oily animal substance deposited in adipose tissue in the body
fatigue [fəˈtiːg] *n.* *(Metall.)* a weakening of a metal due to continuous blows or stress ▶ The series of crashes of Comet aircraft was attributed to metal fatigue.
fault [fɔːlt] *n.* *(Geol.)* a break in the continuity of the layers of rock at or near the surface of the Earth ▶ The notorious San Andreas fault runs through California and threatens San Francisco with periodic earthquakes.
fauna [ˈfɔːnə] *n.* *(Zool.)* the animals found in a specific area ▶ Scientists are making detailed studies of the fauna of the rainforest before it is too late.
fax^1 [fæks] *(abbr.)* **facsimile** [fækˈsɪməlɪ] *n.* *(Elec.)* a system for electronically scanning, transmitting, and reproducing documents by telephone ▶ The two companies exchanged faxes in a matter of minutes although they were hundreds of miles apart.
fax^2 *v.t.* *(Elec.)* to transmit a message by fax ▶ The results of the tests were immediately faxed from the laboratory to the hospital.
fax machine a machine for transmitting and receiving faxed messages
fax number the number by which a transmitter or recipient of a fax is identified
feedback [ˈfiːdˌbæk] *n.* *(Elec.: Mech.)* the return of the output of a circuit or system to the input ▶ The feedback of part of the sound output from the loudspeaker to the microphone caused a high-pitched whistle that hurt one's ears.
feeler [ˈfiːlə] *n.* *(Biol.)* one of a pair of sensitive antennae on an insect's head which enable it to locate food, sense danger, etc.
ferment [fɜˈment] *v.t.* and *v.i.* *(Chem.)* to

effervesce; to be, or cause to be, in a state of fermentation

fermentation [ˌfɜmenˈteɪʃ(ə)n] *n. (Chem.)* the process brought about in certain liquids and other substances by the addition of chemical agents or living organisms, leading to effervescence, giving off of heat, bubbling etc. ▶ Wine is produced from grape juice by fermentation.

ferro- *comb. form* : having to do with a substance containing iron *(Mining)*, or a compound of iron *(Chem.)*

ferro-concrete [ˌferəʊˈkɒŋkrit] concrete strengthened by the addition of iron strips ▶ Ferro-concrete will bear more weight than concrete which has not been reinforced.

ferrous [ˈferəs] *a. (Chem.)* having to do with or containing bivalent iron

fertile [ˈfɜtaɪl] *a. (Biol.: Agric.)* able to bear offspring or fruit ▶ The cows were very fertile and the herd soon grew in size. ▶ The fertile land yielded excellent crops.

fertilize [ˈfɜtəlaɪz] *v.t. (Biol.: Agric.)* to impregnate or make productive

fertilizer [ˈfɜtəˌlaɪzə] *n. (Chem.)* a chemical added to the soil to improve its ability to promote growth ▶ The use of chemical fertilizers can make formerly barren land productive.

fester [ˈfestə] *v.i. (Med.)* (of a wound) to become infected and ulcerate or ooze pus

fever [ˈfivə] *n. (Med.)* any disease that is characterised by raised temperature, high pulse-rate, and loss of energy

fibre [ˈfaɪbə] *n.* a thin thread of material of which animals and plants are constituted, or one manufactured from animal or vegetable tissue ▶ Cotton and wool fibres are used in making clothes.

fibre optics a branch of technology which uses thin fibres of glass or plastic to transmit light ▶ Fibre optics is used in telecommunications and exploratory medicine.

fibreglass [ˈfaɪbəglɑs] *n.* very fine threads of glass, woven into fibre ▶ Fibreglass is used for a number of industrial purposes, one of which is its use as insulating material.

fibrosis [faɪˈbrəʊsɪs] *n. (Med.)* rheumatism of the muscles

fibrous [ˈfaɪbrəs] *a. (Biol.: Med.)* (of tissue) composed of a number of closely knit strands ▶ Scars are made up of fibrous tissue, and are visible because it will not stretch.

filament [ˈfɪləmənt] *n. (Elec.)* a thread of carbon or metal used in electric light bulbs ▶ When the filament snaps, the bulb will no longer light up.

file [faɪl] *n. (Comput.)* a block of data with a name by which it can be accessed ▶ Entries for the dictionary were loaded in files in alphabetical order.

filler [ˈfɪlə] *n. (Build.)* one of a number of materials used for filling cracks and crevices in walls, etc. ▶ Fillers are especially useful in preparation for redecoration of plaster and woodwork.

filter[1] [ˈfɪltə] *n. (Chem.)* an apparatus for straining liquids or gases to remove impurities ▶ Water is passed through a series of filters of sand, charcoal, etc., to make it fit to drink.

filter[2] *v.t. (Chem.)* to purify by the use of filters ▶ The tap-water now has to be filtered before you can drink it.

filtrate [ˈfɪltreɪt] *n. (Chem.)* any liquid that has been passed through a filter

firebreak [ˈfaɪəbreɪk] *n. (Agric.)* a strip of land kept clear of trees and vegetation to prevent the spread of fire

firebrick [ˈfaɪəbrɪk] *n. (Build.)* a brick capable of withstanding great heat, used for building fireplaces

fireclay [ˈfaɪəˌkleɪ] *n. (Build.)* a kind of clay which is capable of withstanding great heat and is used for making firebricks

firedamp [ˈfaɪədæmp] *n. (Chem.: Mining)* an explosive gas which collects in coalmines

fissile [ˈfɪsaɪl] *a. (Phys.)* (of material) capable of being split apart ▶ The core of a nuclear reactor houses the fissile material.

fission [ˈfɪʃ(ə)n] *n. (Phys.: Biol.)* the act of splitting apart ▶ The energy released when an atomic bomb explodes is the result of nuclear fission. ▶ Some primitive

fissionable ['fɪʃ(ə)nəbl] *a. (Phys.)* capable of being split apart by fission

fissiparous [fɪ'sɪpərəs] *a. (Biol.)* having to do with cells that divide into several parts

fit [fɪt] *n. (Med.)* a sudden attack of epilepsy characterized by violent spasms ▶ People who suffer from epileptic fits often do not remember having had them.

flammable ['flæməb(ə)l] *a. (Phys.)* easily ignited ▶ 'Flammable' is now preferred to 'inflammable', which is sometimes wrongly thought to mean 'non-flammable'.

flange [flændʒ] *n. (Mech.)* a projecting rim on a wheel which enables it to be guided along a rail ▶ The wheels of railway carriages have flanges to keep them on the rails.

flap [flæp] *n. (Aer.)* a movable part of the wing of an aircraft that is used to increase lift on take-off and drag on landing ▶ The pilot raised the flaps to the maximum when coming in to land.

flex [fleks] *n. (Elec.)* flexible, insulated wire used for connecting electrical apparatuses to an electricity supply ▶ A length of flex ran along the floor from the wall-plug to the lamp.

float [fləʊt] *n. (Mech.)* the ball of a ballcock that regulates the flow of water from a cistern ▶ When the water in the tank had risen to the right level, the float closed the valve and stopped the flow.

floppy disk [ˌflɒpɪ 'dɪsk] a flexible, magnetic computer disk on which data may be stored

flora ['flɔːrə] *n. (Bot.)* the whole vegetation of an area or historical period ▶ The flora of the island was described in detail in the logbook of the early explorers.

flue [fluː] *n. (Build.: Mech.)* a passage or tube through which smoke and fumes can escape ▶ When a bird nested in the chimney flue, the house was filled with smoke.

fluid¹ ['fluːɪd] *a. (Phys.: Chem.)* (of a substance) composed of particles that move freely in relation to each other, so that it is capable of flowing, like water or a gas

fluid² *n. (Phys.)* a substance that is capable of flowing like water or a gas

fluid drive a system of transmitting power through the movement of oil ▶ Fluid drive is an application of hydraulics

fluorescence [ˌfluəˈres(ə)ns] *n. (Phys.)* a characteristic of certain substances which absorb invisible (short-wave) light and give off visible (longer-wave) light

fluorescent light light emitted by a glass tube coated on the inside with a fluorescent substance when an electric current is passed through it ▶ The harsh fluorescent light showed up every detail of the road.

flush [flʌʃ] *a. (Mech.)* level, not protruding ▶ The head of the screw was flush with the surface of the metal bar.

flutter ['flʌtə] *n. (Radio)* a distortion of sound which may occur in the higher audio frequencies of an audio system

flyover ['flaɪˌəʊvə] *n. (Eng.)* an intersection of two roads at which one is carried over the other by a bridge

flywheel ['flaɪwiːl] *n. (Mech.)* a heavy, rimmed wheel attached to a machine to regulate the speed of the machine by its own inertia ▶ When the flywheel ceased to function, the clock started gaining several hours a day.

focal ['fəʊk(ə)l] *a. (Phys.)* having to do with focus ▶ The distance between the centre of a lens and its focus is called its focal length.

focus ['fəʊkəs] *pl.* **foci** ['fəʊsaɪ] *n. (Phys.)* the point at which rays of light meet after passing through a lens

foetal ['fiːt(ə)l] *a. (Biol.)* having to do with a foetus ▶ The doctor examined the pregnant woman and listened to the foetal heartbeat in her womb.

foetus ['fiːtəs] *n. (Biol.)* the young of a mammal while still inside its mother's uterus

forceps ['fɔːseps] *n. (Med.)* an instrument like a pair of pincers, used for gripping things during operations ▶ It was a difficult birth, and a forceps delivery was necessary.

forensic [fəˈrensɪk] *a.* having to do with the law courts or legal proceedings

forensic medicine the science of medicine in relation to the law ▶ Because of the doubt over the cause of death, it was decided to consult an expert in forensic medicine.

forge [fɔdʒ] *n. (Metall.)* a furnace or hearth for making wrought iron ▶ The flames of the giant forge lit up the sky for miles around.

forklift truck a vehicle which raises and transports heavy objects on mobile steel prongs ▶ The crates in the warehouse were stacked by forklift trucks.

formaldehyde [fɔˈmældɪhaɪd] *n. (Chem.)* a strong-smelling, colourless gas which can be dissolved in water for use as an antiseptic or preserving fluid ▶ The biological specimens are preserved in formaldehyde in the school laboratories.

formalin [ˈfɔməlɪn] *n. (Chem.: Med.)* a solution of formaldehyde used as an antiseptic

formula [ˈfɔmjʊlə] *pl.* **formulas** or **formulae** *n. (Chem.: Maths.)* an expression of the chemical composition of a substance or of an algebraic equation by means of letters and numbers ▶ Everyone knows that the formula H_2O, denoting a substance composed of two parts hydrogen (H) to one part oxygen (O), signifies water.

Fortran (FORTRAN) [ˈfɔtræn] *n. (Comput.)* a high-level computer language, used especially in programs of mathematics and science, in which formulae are common.

fossil [ˈfɒsl] *n. (Geol.)* the body of a plant, animal or other organism preserved in Earth's crust, e.g. in the rocks ▶ Fossils of huge dinosaurs were discovered almost complete in the Gobi desert in Mongolia.

fossil fuel a naturally-occuring fuel, such as coal, which originates from vegetation ▶ The burning of fossil fuels is a major cause of pollution of the atmosphere.

fossilize [ˈfɒsəlaɪz] *v.t.* and *v.i.* (Geol.) to convert into or be converted into a fossil

foundry [ˈfaʊndrɪ] *n. (Metall.)* a building where metals are cast

fraction [ˈfrækʃ(ə)n] *n. (Maths.)* the expression of one or more parts of a unit ▶ Probably the most commonly used fractions are $\frac{1}{4}$, $\frac{1}{2}$, and $\frac{3}{4}$.

fracture¹ [ˈfræktʃə] *v.t.* and *v.i. (Med.)* (of a bone) to break

fracture² *n. (Med.)* (of a bone) a break ▶ When only the bone is broken, it is a simple fracture; but if the surrounding tissue is also damaged, it is a compound fracture.

freeze [friz] *v.t.* and *v.i. (Phys.)* (of a fluid) to turn, or cause to turn, into a solid as a result of the action of cold ▶ The water in the well had frozen solid.

freezer [ˈfrizə] *n.* a cabinet or compartment for long-term freezing of perishable goods, especially foodstuffs ▶ Food kept in a freezer will remain eatable for long periods.

freezing-point [ˈfrizɪŋˌpɔɪnt] the point at which a fluid freezes ▶ The freezing-point of water is 32°F or 0°C.

frequency [ˈfrikwənsɪ] *n. (Elec.)* the speed of variations of alternating currents etc.; the rate at which something vibrates, e.g. the human vocal cords ▶ If a sound wave vibrates quickly, it has high frequency and a high note may be heard. ▶ The radio was not audible so we tuned into another frequency.

frequency modulation the variation of frequency of a radio carrier wave according to the characteristics of the sound being broadcast, e.g. speech or music

friable [ˈfraɪəbl] *a. (Hort.)* (of soil) readily crumbled, easy to dig ▶ Digging compost into the heavy soil made it more friable.

friction [ˈfrɪkʃ(ə)n] *n. (Phys.)* the resistance any body meets when rubbing against another body ▶ The friction created when moving parts of a machine are in contact generates heat, which may be cooled by a constant stream of lubricating oil.

fuel [ˈfjʊəl] *n. (Chem.)* material which can be burned or used in a nuclear reactor ▶ Domestic fuels include wood, coal, gas and oil.

fulcrum [ˈfʊlkrəm] *n.* *(Phys.: Mech.)* the fixed point on which the bar of a lever rests or about which it turns ▶ The fulcrum of a common see-saw is in the middle.

fume [fjum] *n.* *(Chem.)* a foul-smelling and frequently toxic smoke ▶ The fumes from the leaking exhaust pipe gradually overcame the driver and he lost consciousness.

fumigate [ˈfjumɪgeɪt] *v.t.* *(Chem.: Med.)* to disinfect an area by means of chemical smoke as a hygienic measure ▶ When the epidemic had subsided, the huts in the infected area had to be thoroughly fumigated before they could be reoccupied.

function [ˈfʌŋkʃ(ə)n] *n.* *(Maths.)* a value which varies in accordance with another value ▶ In the equation $a = 3b$, a is a function of b, because if a changes, so does b.

fungicide [ˈfʌndʒɪsaɪd] *n.* *(Chem.)* a chemical substance which kills fungi

fungus [ˈfʌŋgəs] *pl.* **fungi** [ˈfʌndʒɪ] *n.* *(Bot.)* a mushroom or mould-like growth which has no chlorophyll and lives on organic matter ▶ The rotting logs were covered with fungi.

funicular [fjʊˈnɪkjʊlə] *a.* *(Mech.)* (of a railway) in which the carriage is pulled up a mountain by an engine stationed at the top and a moving steel cable ▶ The tourists were able to admire the view by taking the funicular railway to the summit of the mountain.

funnel [ˈfʌn(ə)l] *n.* *(Mech.)* a conical vessel, usually with a tube at the bottom, for transferring liquids and other substances from one container to another.

furnace [ˈfɜnɪs] *n.* *(Metall.)* a chamber or other structure where fuel is burned in order to generate intense heat for melting iron ore, etc.

fuse[1] [fjuz] *n.* *(Mech.)* a device for detonating a bomb or shell

fuse[2] *v.t.* *(Mech.)* to arm a bomb or shell with a fuse preparatory to exploding or firing it

fuse[3] *n.* *(Elec.)* a device containing a piece of fine wire inserted into an electrical system, which melts and thus breaks the circuit when it is overloaded ▶ When too many electrical appliances are switched on at the same time, the fuse blows and the circuit is broken.

fuse[4] *v.t.* and *v.i.* *(Metall.)* to unite by melting together ▶ In the intense heat of the fire, the metal parts had fused together and could not be separated without being broken.

fuselage [ˈfjuzɪlɑʒ] *n.* *(Aer.)* the main body of an aircraft or rocket ▶ The wings and tail were broken off in the crash, but the fuselage remained virtually intact.

fusion[1] [ˈfjuʒ(ə)n] *(Metall.)* the act of melting something or rendering it liquid by the action of heat

fusion[2] *n.* *(Phys.)* the combination at very high temperature of atomic nuclei of hydrogen or deuterium to form helium nuclei and liberate nuclear energy, as in the explosion of a hydrogen bomb

g

galaxy [ˈgæləksɪ] *n.* (*Astron.*) any large cluster of stars including or beyond the Milky Way ▶ As techniques improve, astronomers are still discovering hitherto unknown galaxies.

galvanic [gælˈvænɪk] *a.* (*Phys.: Chem.*) having to do with the production of electricity by chemical means ▶ The action of an acid on a metal produces galvanic electricity.

galvanize [ˈgælvənaɪz] *v.t.* (*Metall.*) to plate iron with zinc by an electrical process to protect it from moisture ▶ Nails for outside use are galvanized so that they do not rust.

galvanometer [gælvəˈnɒmɪtə] *n.* (*Elec.*) a delicate apparatus for determining the existence, direction, and intensity of electrical currents

gamma [ˈgæmə] *n.* (*Phys.*) a unit for measuring the strength of a magnetic field

gamma globulin any of a group of proteins carried in blood and serum and including antibodies ▶ The European tourist was given an injection of gamma globulin before going on holiday in Africa.

gamma rays (*Phys.: Med.*) short-wavelength rays emitted by radioactive substances and used in the treatment of cancer ▶ Uncontrolled exposure to gamma rays can be extremely dangerous.

gangrene [ˈgæŋgrin] *n.* (*Med.*) cessation of life in a part of the body, because the blood supply has stopped, leading to decay ▶ Wounds not treated immediately may develop fatal gangrene.

gas [gæs] *n.* (*Chem.*) a substance that is neither liquid nor solid ▶ A gas may have no colour or smell.

gaseous [ˈgeɪsɪəs] *a.* (*Chem.*) having to do with gas

gasket [gæskɪt] *n.* (*Eng.*) a strip of material used on pipes or engine cylinders to make them air-tight or water-tight ▶ A broken gasket in the car engine will lead to loss of power.

gastric [ˈgæstrɪk] *a.* (*Med.*) having to do with the stomach ▶ He was suffering from a gastric ulcer that gave him considerable pain.

gastroenteritis [ˌgæstrəʊˌentəˈraɪtɪs] *n.* (*Med.*) inflammation of the stomach and intestines

gauge^1 [geɪdʒ] *n.* (*Meas.*) a device for measuring rainfall, wind velocity, thickness of wire etc. ▶ The rain gauge showed an unusually heavy rainfall for the time of year.

gauge^2 *n.* (*Meas.*) a standard size or thickness of wire, screws, and other metal parts ▶ It is important to use fuse-wire of the right gauge to mend a particular fuse.

gauge^3 *n.* (*Eng.*) the distance between the rails of a railway track ▶ Trains travelling over great distances normally use lines with a broad gauge.

gear [gɪə] *n.* (*Eng.*) a set of toothed wheels interacting with one another to control the speed or direction of a machine ▶ Most modern car engines have five forward gears.

gear-box [ˈgɪəbɒks] *n.* the casing in which the gears are enclosed in a motor vehicle or other machine

gear-lever [ˈgɪəlivə] *n.* a device for selecting or changing from one gear to another in a car engine etc. ▶ Move the gear-lever forward to engage third gear.

Geiger [ˈgaɪgə] **counter** a device for detecting and counting particles emitted by radioactive materials ▶ After the accident,

Geiger counters showed the presence of radioactive particles in the area surrounding the nuclear installation.

gene [dʒiːn] *n.* *(Biol.)* one of the units of heredity which determine characteristics passed on from one generation to another

generate [ˈdʒenəreɪt] *v.t.* *(Elec.)* to cause or bring into being ▶ Electricity is generated at coal-burning, gas-burning, and nuclear power stations.

generator [ˈdʒenəreɪtə] *n.* *(Elec.)* an apparatus for producing gas, electricity, steam, etc.

genetic [dʒəˈnetɪk] *a.* *(Biol.)* having to do with the origin or creation of something

genetic engineering the artificial alteration of the genes of an organism in order to control its transmission of hereditary characteristics ▶ Recently developed techniques in genetic engineering have raised controversial ethical questions.

geneticist [dʒəˈnetɪsɪst] *n.* *(Biol.)* an expert in genetics

genetics [dʒəˈnetɪks] *n. sg.* *(Biol.)* the study of heredity ▶ Recent advances in genetics have clarified some of the processes by which a person's character is formed.

genus [ˈdʒiːnəs] *pl.* **genera** [ˈdʒenərə] n. (Biol.) a group or class of plants or animals differentiated from all others by certain common characteristics ▶ Genus comes below family in the taxonomic classification.

geological [dʒiəˈlɒdʒɪk(ə)l] *a.* having to do with geology

geologist [dʒiˈɒlədʒɪst] *n.* an expert in geology

geology [dʒiˈɒlədʒɪ] *n.* the scientific study of rocks and soils and of their history

geometrical [dʒiəˈmetrɪk(ə)l] *a.* *(Maths.)* having to do with geometry

geometrical progression a sequence of numbers in which each differs from the next one by a common ratio ▶ The sequence 2-6-18-54 is a geometrical progression.

geometry [dʒiˈɒmətrɪ] *n.* *(Maths.)* a branch of mathematics concerned with lines, angles and shapes ▶ The science of geometry was known to the ancient Greeks, who used it in designing the famous classical buildings.

geophysics [ˌdʒiəˈfɪzɪks] *n. sg.* *(Geol.)* the science that deals with the physical characteristics of the Earth ▶ An important branch of geophysics is meteorology.

geriatrics [ˌdʒerɪˈætrɪks] *n. sg.* *(Med.)* the branch of medicine dealing with old age and its diseases ▶ As life expectancy increases, the science of geriatrics becomes more important.

germ [dʒɜːm] *n.* *(Biol.)* an extremely small living organism that causes diseases in animals or plants

germicide [ˈdʒɜːmɪsaɪd] *n.* *(Chem.)* a substance used for killing germs ▶ The systematic use of germicides is a part of hygiene.

gerontology [ˌdʒerɒnˈtɒlədʒɪ] *n.* *(Med.)* the science dealing with deterioration and decay in the aged

gestate [dʒeˈsteɪt] *v.i.* *(Biol.)* to develop in the mother's uterus

gestation [dʒeˈsteɪʃ(ə)n] *n.* *(Biol.)* the development of an embryo or baby in the mother's uterus ▶ The period of gestation varies from one mammal to another.

geyser [ˈgiːzə] [ˈgaɪzə] *n.* *(Geol.)* a hot spring throwing up a column of water at intervals ▶ Geysers are a natural source of hot water in New Zealand, the USA and Iceland.

girder [ˈgɜːdə] *n.* *(Eng.)* a main beam, often assembled of metal parts, spanning a building or structure ▶ A series of iron girders were supported on stone pillars on either side of the gorge and a road-bridge suspended between them.

glacial [ˈgleɪsɪəl] *a.* *(Geol.)* resulting from the effect of glaciers

glacier [ˈglæsɪə] *n.* *(Geol.)* a stream-like mass of ice formed by snow at high altitudes and descending slowly to lower regions ▶ The gradual movement of massive glaciers scooped out huge valleys.

gland [glænd] *n.* *(Med.)* an organ that secretes certain constituents of the blood, either for use by the body or for excretion ▶ Poor functioning of the thyroid gland can lead to obesity.

glandular [ˈglændjʊlə] *a.* *(Med.)* having to do

with glands ▶ Glandular fever is an infection leading to enlargement of the lymph glands.

glaucoma [glɔˈkəʊmə] *n. (Med.)* a disease of the eye in which increased pressure inside the eyeball damages the optic disc and may impair vision

global [ˈgləʊb(əl)] *a. (Geog.)* having to do with the whole of the Earth ▶ Full-scale nuclear war could result in devastation on a global scale.

global warming the rise in temperature of the Earth's atmosphere due to the effect of excessive burning of fossil fuels

globe [gləʊb] *n. (Geog.)* a sphere representing the Earth, on which a map is painted showing the land and sea

glucose [ˈglukəʊz] *n. (Chem.)* a substance less sweet than cane-sugar, obtained from various fruits and used in the manufacture of foodstuffs ▶ The list of ingredients on a food packet should show whether the contents include glucose.

glycerine [ˈglɪsərin] *n. (Chem.)* a sweet, colourless liquid obtained from animal and vegetable fats and oils and used in the manufacture of soap, medicines and sweets

gonorrhoea [ˌgɒnəˈriə] *n. (Med.)* a venereal disease affecting the urethra ▶ Gonorrhoea is now easily cured by drug treatment and is becoming increasingly rare.

governor [ˈgʌv(ə)nə] *n. (Eng.)* a device for regulating the flow of a fluid or gas in an engine ▶ A governor was installed in the car engine to prevent it exceeding a set speed.

gradation [grəˈdeɪʃ(ə)n] *n.* (usually *pl.*) a stage or degree in development or gradual change ▶ A trained ear can distinguish gradations of sound in the pronunciation of common vowels.

graduate [ˈgrædjʊeɪt] *v.t.* to mark something, such as a ruler, with regular divisions

graduated [ˈgrædjʊeɪtɪd] *a. (Meas.)* marked with regular divisions ▶ The sizes of the samples were measured on a graduated scale.

graft [grɑft] *v.t. (Med.: Hort.)* to implant living tissue from one plant or animal to another so that the tissue continues to grow ▶ Many fruit trees are grafted on to the stocks of other, more hardy trees.

gram (gramme) [græm] *n. (Meas.)* the standard unit of weight in the metric system (about 0·04 oz.)

graph [grɑf] *n. (Maths.)* a diagram showing by lines and curves how two or more quantities are related ▶ The increase in the number of products sold over a certain number of years was clearly shown on the graph.

graph paper [ˈgrɑf ˌpeɪpə] paper ruled in small squares to facilitate the drawing of graphs

gravitate [ˈgrævɪteɪt] *v.i. (Phys.)* to move downwards as though pulled by gravity ▶ The heavier elements in the solution gravitated naturally to the bottom of the flask.

gravity [ˈgrævətɪ] *n. (Phys.)* the force causing bodies to move downwards towards the centre of the Earth, making them appear to have weight

grease gun [ˈgris gʌn] a device for forcing grease or lubricating oil into machinery

greenhouse [ˈgrinhaʊs] *n. (Agric.)* a glasshouse for cultivating and preserving tender plants ▶ Tomatoes are grown in greenhouses to protect them from frost.

greenhouse effect [ˈgrinhaʊs Iˌfekt] the increased temperature of the Earth caused by the atmosphere behaving as the glass of a greenhouse does ▶ Enviromentalists are becoming very worried at the increasing influence of the greenhouse effect on climate.

grid[1] [grid] *n. (Mech.)* a grating of parallel bars ▶ The grid of the barbecue was glowing.

grid[2] *n. (Eng.: Elec.)* a system of distributing electricity over large areas ▶ Current was maintained after the storm by distribution through the national grid.

groundspeed [ˈgraʊn(d)ˌspid] *n. (Aer.)* the speed of an aircraft relative to the ground ▶ Because of a strong headwind, the aircraft's groundspeed was considerably less than its airspeed.

grub [grʌb] *n. (Biol.)* the larva of an insect, especially of a bee or wasp ▶ A grub has no legs and a head that is no wider than the rest of its body.

grub screw [ˈgrʌb ˌskru] a small, headless screw which does not protrude above the surface into which it is screwed

gynaecology [ˌgaɪnɪˈkɒlədʒɪ] *n. (Med.)* the science dealing with functions and diseases peculiar to women

gyroscope [ˈdʒaɪərəˌskəʊp] *n. (Mech.)* a flywheel, rotating at a very high speed, which is used to maintain a steady direction ▶ Gyroscopes play an important part in controlling and stabilizing compasses in ships and aircraft.

h

ha *(abbr.)* hectare
habitat [ˈhæbɪtæt] *n.* *(Ecol.)* the natural abode or locality in which a plant or animal is found ▶ Draining the marshes destroyed the natural habitats of several rare species of birds.
hack (into) [hæk] *v.t.* *(Comput.)* to gain illegal access to someone else's program ▶ The thieves hacked into the bank's computer program and transferred millions of dollars to their own accounts.
hacker *n.* *(Comput.)* someone who gains illegal access to a computer program
haema-, haemo- *comb. form* blood
haematology [ˌhiməˈtɒlədʒɪ] *n.* *(Med.)* the scientific study of diseases of the blood
haemoglobin [ˌhiməˈgləʊbɪn] *n.* *(Med.)* the colouring matter of the red corpuscles of the blood
haemophilia [ˌhiməˈfɪlɪə] *n.* *(Med.)* an inherited disease of the blood which prevents it from clotting ▶ Sufferers from haemophilia have to be very careful to avoid wounds that bleed.
haemophiliac [ˌhiməˈfɪlɪək] *n.* *(Med.)* someone suffering from haemophilia
haemorrhage [ˈhemərɪdʒ] *n.* *(Med.)* abnormal discharge of blood ▶ Excessive loss of blood from a series of haemorrhages weakened his chances of survival.
halitosis [ˌhælɪˈtəʊsɪs] *n.* *(Med.)* bad-smelling breath ▶ Halitosis often results from tooth decay.
hallucinate [həˈluːsɪneɪt] *v.i.* *(Psych.)* to wander in the mind, especially seeing things that are not actually there ▶ An overdose of drugs caused her to hallucinate and she began to see things that no one else could see.
hallucinogen [ˌhæluˈsɪnədʒ(ə)n] *n.* *(Chem.:* *Psych.)* a drug that induces hallucination
halogen [ˈhælədʒen] *n.* *(Chem.)* one of a group of elements that can combine with a metal to form a salt
hangar [ˈhæŋə] *n.* *(Aer.)* a large shed in which aircraft are kept
hang-glider [ˈhæŋˌglaɪdə] *n.* a type of large kite controlled by a person suspended underneath it in a harness
hard drug a dangerous and addictive drug ▶ Cocaine and heroin are hard drugs.
hard water water with a mineral content that inhibits the formation of soapsuds
hardware [ˈhɑdweə] *n.* *(Comput.)* the actual machinery of a computer, etc., as opposed to programs (software) used with it ▶ No matter how sophisticated the hardware may be, it is useless without the right software.
harmonics [hɑˈmɒnɪks] *n.* *(Phys.)* the study of the physical characteristics of musical notes
hatch [hætʃ] *n.* *(Naut.)* a trapdoor to cover a hatchway ▶ The hatches were fastened down in preparation for a storm.
hatchback [ˈhætʃˌbæk] a car with a door at the back that opens upward ▶ Hatchbacks are popular cars with owners of dogs.
hatchway [ˈhætʃˌweɪ] a large opening in the deck of a ship through which cargo may be lowered or raised
hawser [ˈhɔzə] *n.* *(Naut.)* a cable used for tying up ships at the quay
head[1] [hed] *n.* *(Elec.)* the device on a tape recorder or video recorder that can record, play back, or erase recorded sound or visual images ▶ To get the best quality reproduction you should occasionally clean the heads of your tape or video recorder.

head² *pl.* **head** *n. (Meas.)* a single one (as of cattle) ▶ The herd was increased to a thousand head.

head³ *n. (Eng.)* pressure of water, steam, etc. ▶ Before leaving harbour, the captain asked the chief engineer to build up a good head of steam.

headroom [ˈhedˌrʊm] *n. (Eng.)* sufficient space for the head of a vehicle etc. ▶ The bridges over the motorway must leave enough headroom for large trucks to pass underneath them.

headwaters [ˈhedˌwɔtəz] *n. pl. (Geog.)* the upper part of a stream, usually flowing directly from its source ▶ Explorers searched for many years to find the headwaters of the Nile.

headwind [ˈhedˌwɪnd] *n. (Aer.: Meteor.)* a wind blowing from the direction in which a vehicle is travelling ▶ The aircraft encountered a strong headwind and arrived late.

heavenly body any sun, star, planet, or other mass of matter, apart from the Earth

heavens [ˈhev(ə)nz] *n. pl. (Astron.)* the sky or atmosphere enveloping the Earth

heavy duty (of machinery, vehicles) designed to withstand exceptionally heavy use

heavy industry industry that makes heavy machines or produces heavy materials, such as ships and coal ▶ Many of the old heavy industry establishments have now been demolished and replaced by light industries, manufacturing electronic parts, etc.

hectare [ˈhekteə] *n. (Meas.)* a measure of area equal to 100,000 square metres (2·471 acres) ▶ Many people still have difficulty in converting hectares into acres.

-hedron *comb. form (Maths.)* (of a solid figure) having the stated number of sides ▶ A tetrahedron has four sides.

helical [ˈhelɪk(ə)l] *a. (Mech.)* like a helix, spiral ▶ In helical gears, the teeth of the gear-wheels are set at an angle.

heli(o)- *comb. form (Astron.: Phys.)* having to do with the sun or the rays of the sun

heliograph [ˈhiliə(ʊ)ɡræf] *n. (Phys.)* an apparatus for signalling by reflecting flashes of sunlight ▶ News of the victory was flashed by heliograph from hill to hill.

heliport [ˈhelɪpɔt] *n. (Aer.)* an airport for helicopters

helium [ˈhiliəm] *n. (Chem.)* a light gas which has no smell and will not burn ▶ Helium-filled balloons are widely used in meteorology.

helix [ˈhilɪks] *pl.* **helices** [ˈhilɪsiz] *n. (Maths.)* a spiral line, as in a coil of wire or rope

hemi- *comb. form* having to do with one-half of anything

hemiplegia [ˌhemɪˈplidʒɪə] *n. (Med.)* paralysis of one side of the body ▶ Hemiplegia is frequently the result of a stroke.

hemisphere [ˈhemɪsfɪə] *n. (Geog.)* half a sphere ▶ The Earth's globe is notionally divided by the equator into northern and southern hemispheres.

hepatitis [ˌhepəˈtaɪtɪs] *n. (Med.)* inflammation of the liver ▶ They were given injections to protect them from hepatitis A. ▶ Injections were not needed against hepatitis B or C.

herbarium [hɜˈbeərɪəm] *n. (Bot.)* a systematic collection of dried plants, or a place for preserving them ▶ The herbarium at Kew Gardens in London is an important research centre.

herbicide [ˈhɜbɪsaɪd] *n. (Chem.: Agric.)* a chemical substance which destroys plants ▶ The indiscriminate use of herbicides to protect crops may do great damage to the environment.

herbivore [ˈhɜbɪvɔ] *n. (Zool.)* an animal that feeds on plants or grass ▶ The largest existing herbivore is the elephant.

heredity [həˈredətɪ] *n. (Biol.)* the passing on of personality or behavioural characteristics by parents to their progeny, or the way in which progeny resemble their parents

hermaphrodite [hɜˈmæfrədaɪt] *n. (Biol.)* an animal or plant combining within itself both male and female sexual organs ▶ Hermaphrodites reproduce themselves independently.

hermetically [hɜˈmetɪk(ə)lɪ] *adv. (Phys.)*

excluding all air ▶ The bottles were hermetically sealed to prevent the possibility of contamination from the atmosphere.

hernia [ˈhɜnɪə] n. (Med.) the protrusion of an organ of the body through a weak place in the covering wall, most commonly when the bowel is pushed through the wall of the abdomen ▶ The swelling in his groin was due to a hernia caused by lifting heavy loads.

heroin [ˈherəʊɪn] n. (Chem.: Med.) a habit-forming drug derived from morphine ▶ She became addicted to heroin and found it impossible to break the habit.

herpes [ˈhɜpiz] n. (Med.) a viral infection of the skin ▶ The virus which causes coldsores, especially in children, is a common type of herpes.

herpetic [hɜˈpetɪk] a. (Med.) having to do with herpes

hertz [hɜts] n. (Radio) a unit of frequency of radio waves, expressed in number of cycles per second ▶ 10,000 cycles per second = 10,000 hertz (or 10 kilohertz).

hetero- comb. form different, various, other

heterogeneous [ˌhetərə(ʊ)ˈdʒɪnɪəs] a. diverse in character ▶ The audience was a heterogeneous collection of people of all ages and backgrounds.

heterosexual [ˌheterə(ʊ)ˈseksjʊəl] a. (Biol.) attracted towards the opposite sex

hexagon [ˈheksəgən] n. (Maths.) a plane figure with six sides

hi-fi (abbr.) a. (Phys.) high-fidelity (having to do with equipment for reproducing sound with very little distortion)

high frequency any frequency of alternating current above the audible range

high-tech. [ˌhaɪˈtek] (abbr.) high technology

high-tension relating to a high and steady electric current ▶ High tension cables carried the current directly from the generator to the factory.

HIV (abbr.) Human Immunodeficiency Virus ▶ The HIV virus usually results in AIDS.

hod [hɒd] n. (Build.) a wooden holder, shaped like a trough on a pole, for carrying bricks

hold [hɒʊld] n. (Naut.) the space inside a ship where the cargo is stored ▶ The hatches were raised and the containers of cargo lowered into the hold.

holding pattern [ˈhɒʊldɪŋ ˌpætən] the course an aircraft is requested to follow while awaiting permission to land ▶ Owing to the traffic congestion at London's Heathrow airport, the plane was put into a holding pattern for almost half an hour.

home (in, on) [həʊm] v.i. (Eng.) to be guided automatically to a certain target ▶ The ground-to-air missile homed in on the approaching aircraft.

homeo- comb. form similar

homeopath [ˈhɒmɪə(ʊ)pæθ] n. (Med.) a specialist in homeopathy

homeopathy [ˌhɒmɪˈɒpəθɪ] n. (Med.) the treatment of diseases by giving small doses of drugs that in healthy people produce the sort of symptoms they are designed to cure

homing device [ˈhəʊmɪŋ dɪˌvaɪs] an electronically-controlled device fitted to a missile which enables it to home in on a specific target

homing instinct [ˈhəʊmɪŋ ˌɪnstɪŋkt] the ability of certain animals and birds to find their way home

homo- comb. form likeness, sameness

homogeneous [ˌhɒmə(ʊ)ˈdʒɪnɪəs] a. having parts which are the same or similar ▶ It is easier to learn in a homogeneous group than in a collection of people who are widely different in level.

homosexual [ˌhɒmə(ʊ)ˈseksjʊəl] a. (Biol.) attracted toward members of the same sex

hone¹ [həʊm] v.t. to sharpen a blade on a hone

hone² n. a stone on which blades of knives etc. are sharpened

hopper [ˈhɒpə] n. (Eng.) a large funnel through which substances such as grain, sand or coal can be passed into containers, vehicles etc.

horizontal [ˌhɒrɪzˈɒnt(ə)l] n. (Maths.) a flat or level line

hormone [ˈhɔːməʊn] n. (Biol.: Chem.) a secretion from an internal gland which activates or stimulates an organ of the body ▶ Adrenalin, a hormone produced at times of sudden fear or anger, raises the blood pressure and stimulates the activity of the heart.

horticulture [ˌhɔtɪˈkʌltʃə] *n. (Agric.)* the science and practice of cultivating gardens

hospice [ˈhɒspɪs] *n. (Med.)* a nursing home or hospital, mainly for the terminally ill

host[1] [həʊst] *n. (Biol.)* a plant or animal on which a parasite lives ▶ Some kinds of trees are often hosts to such parasites as mistletoe or ivy.

host[2] *n. (Med.: Biol.)* an organism into which an organ or tissue from another organism is grafted or transplanted ▶ The host body sometimes rejects a transplanted organ.

hothouse [ˈhɒthaʊs] *n. (Hort.)* a heated building for growing or preserving plants or animals in temperatures higher than those outside ▶ A number of reptiles and tropical plants can be observed in hothouses in the Wild Life park.

hovercraft [ˈhɒvəkrɑft] *n. (Aer.)* an aircraft able to move at low level over land or water supported on a cushion of air created by fans

hoverport [ˈhɒvəpɔt] *n. (Aer.)* a place where passengers enter or leave hovercraft

hull [hʌl] *n. (Naut.)* the main body of a ship ▶ The hull of the ship was holed below the waterline and the sea began to pour in.

humid [ˈhjumɪd] *a. (Meteor.)* (of the atmosphere) warm and damp

humidifier [hjuˈmɪdɪfaɪə] *n. (Mech.)* a piece of apparatus that prevents the air in a room from becoming too dry ▶ A humidifier may add drops of water to the air circulation system.

humidity [hjuˈmɪdɪtɪ] *n. (Meteor.)* a measure of the amount of moisture in the atmosphere

humus [ˈhjuməs] *n. (Hort.)* material composed of rotted leaves and plants ▶ The soil was improved by digging in humus.

husbandry [ˈhʌzbəndrɪ] *n. (Agric.)* the systematic or scientific practice of a branch of agriculture or farming ▶ Before taking up cattle farming, he consulted an expert in animal husbandry.

hybrid [ˈhaɪbrɪd] *n. or a. (Biol.)* a plant or animal which has been bred from two distinct species or varieties ▶ Hybrid roses can combine the desired colour and perfume.

hybridization [ˌhaɪbrɪdaɪˈzeɪʃ(ə)n] *n. (Biol.)* the process or result of cross-fertilization or interbreeding of plants or animals

hydr-, hydro- *comb. form* having to do with water

hydrant [ˈhaɪdrənt] *n. (Eng.)* a spout or discharge pipe, usually with a nozzle for attaching a hose, connected with a water main ▶ The firemen attached their hoses to the hydrants in the street and quickly doused the fire.

hydrate [ˈhaɪdreɪt] *n. (Chem.)* a compound of water with an element or another compound

hydraulic [haɪˈdrɒlɪk] *a. (Eng.)* having to do with the operation of equipment by pressure of water or another fluid ▶ When the brake pedal of a car is depressed, hydraulic fluid is forced through small pipes and activates the brake mechanism.

hydraulics [haɪˈdrɒlɪks] *n. sg. (Eng.)* the branch of mechanical engineering concerned with water and other fluids in movement

hydrocarbon [ˌhaɪdrə(ʊ)ˈkɑbən] *n. (Chem.)* a compound of hydrogen and carbon

hydrochloric [ˌhaɪdrə(ʊ)ˈklɒrɪk] *a. (Chem.)* containing chlorine and hydrogen ▶ Hydrochloric acid is used in a number of industrial processes.

hydroelectric [ˌhaɪdrə(ʊ)ɪˈlektrɪk] *a. (Eng.: Elec.)* having to do with the production of electricity by water power ▶ Hydroelectric stations are often built near river dams.

hydrofoil [ˈhaɪdrə(ʊ)fɔɪl] *n. (Naut.)* a fast vessel with one or more pairs of vanes attached to its hull which lift it out of the water at speed

hydrogen [ˈhaɪdrədʒən] *n. (Chem.)* an invisible, flammable gaseous element ▶ Hydrogen is the lightest of all known substances. ▶ Hydrogen combines with oxygen to form water.

hydrogen bomb [ˈhaɪdrədʒən ˌbɒm] an exceedingly powerful bomb in which an immense release of energy is obtained by converting hydrogen into helium by fusion

hydrography [haɪˈdrɒgrəfɪ] *n.* *(Geog.)* the scientific study and mapping of seas and rivers

hydrology [haɪˈdrɒlədʒɪ] *n.* *(Phys.: Meteor.)* the scientific study of water and its properties on the Earth's surface and in its atmosphere

hydrolysis [haɪˈdrɒləsɪs] *n.* *(Chem.)* the formation of an acid and a base by the action of water ▶ The hydrolysis of fats is important in the manufacture of soap.

hydrophobia [ˌhaɪdrə(ʊ)ˈfəʊbɪə] *n.* *(Psych.)* an unnatural dread of water ▶ Hydrophobia is a symptom of rabies.

hydroplane [ˈhaɪdrə(ʊ)ˌpleɪn] *n.* *(Naut.)* a flat-bottomed speedboat which partly rises out of the water at high speed

hydroscope [ˈhaɪdrə(ʊ)ˌskəʊp] *n.* *(Eng.)* an instrument for viewing under water

hydrosphere [ˈhaɪdrə(ʊ)ˌsfɪə] *n.* *(Phys.)* the watery layer converting much of the surface of the Earth

hydrotherapy [ˌhaɪdrə(ʊ)ˈθerəpɪ] *n.* *(Med.)* the treatment of diseases by the use of water

hygiene [ˈhaɪdʒin] *n.* *(Med.)* the science of the prevention of disease by improved sanitation and cleanliness ▶ The restaurant was closed down after the inspection revealed a poor standard of hygiene in the kitchen.

hygienics [haɪˈdʒɪnɪks] *n. sg.* *(Med.)* the principles of hygiene

hygrometer [haɪˈgrɒmɪtə] *n.* *(Meteor.)* an instrument for measuring the humidity of the air

hygroscope [ˈhaɪgrəˌskəʊp] *n.* *(Meteor.)* a device that indicates changes in humidity

hyper- *comb. form* above, beyond, in excess

hyperacidity [ˌhaɪpərəˈsɪdɪtɪ] *n.* *(Med.)* the state of containing more acid than normal ▶ Hyperacidity of the stomach is often related to indigestion.

hyperaesthesia [ˌhaɪpərɪsˈθizɪə] *n.* *(Med.)* abnormal sensitivity to heat, cold, pain, sound etc.

hypertension [ˌhaɪpəˈtenʃ(ə)n] *n.* *(Med.)* abnormally high blood pressure

hyperthermia [ˌhaɪpəˈθɜmɪə] *n.* *(Med.)* abnormally high body temperature

hypertrophy [haɪˈpɜtrəfɪ] *n.* *(Med.)* excessive development or enlargement of an organ of the body

hyperventilation [ˌhaɪpəventɪˈleɪʃ(ə)n] *n.* *(Med.)* excessive breathing, causing abnormal loss of carbon dioxide in the blood ▶ A frequent cause of hyperventilation is over-anxiety.

hypno- *comb. form* having to do with sleep

hypnosis [hɪpˈnəʊsɪs] *n.* *(Psych.)* a semiconscious state in which a person's attitudes, thoughts, and actions can be influenced by statements and suggestions made by another person ▶ Use is made of hypnosis in the treatment of various phobias and harmful habits, like smoking.

hypnotist [ˈhɪpnətɪst] *n.* *(Psych.)* a person who induces a state of hypnotism

hypnotize [ˈhɪpnətaɪz] *v.t.* *(Psych.)* to induce a state of hypnosis in a person

hypo- *comb. form* below, under

hypochondria [ˌhaɪpə(ʊ)ˈkɒndrɪə] *n.* *(Psych.)* excessive and unnecessary worrying about one's health

hypochondriac [ˌhaɪpə(ʊ)ˈkɒndrɪæk] *n.* *(Psych.)* a person who suffers from hypochondria ▶ Although he was perfectly healthy, he became a hypochondriac and spent a fortune on medicines and pills.

hypodermic [ˌhaɪpə(ʊ)ˈdɜmɪk] *a.* *(Med.)* having to do with parts lying under the skin

hypodermic needle the hollow needle of a hypodermic syringe

hypodermic syringe a small syringe with a hollow needle for giving hypodermic injections ▶ Addicts using unsterile hypodermic syringes risk contracting the HIV virus.

hypotension [ˌhaɪpə(ʊ)ˈtenʃ(ə)n] *n.* *(Med.)*

abnormally low blood pressure, possibly resulting from shock

hypotenuse [haɪˈpɒtənjuːz] *n.* *(Maths.)* the side of a rightangled triangle opposite the right angle ▶ Pythagoras established that the square on the hypotenuse equals the sum of the squares on the two other sides.

hypothermia [ˌhaɪpəʊˈθɜːmɪə] *n.* *(Med.)* abnormally low body temperature ▶ Every winter a number of elderly people die of hypothermia because of lack of adequate heating in their homes.

hypothesis [haɪˈpɒθəsɪs] *n.* a proposition adopted for the sake of argument ▶ Astronomers have put forward various hypotheses to explain the creation of the universe.

hypothetical [ˌhaɪpəˈθetɪk(ə)l] *a.* founded on a hypothesis ▶ The lecturer cited a hypothetical case to illustrate his argument.

hysterectomy [ˌhɪstəˈrektəmɪ] *n.* *(Med.)* a surgical operation for removing the uterus

hysteria [hɪsˈtɪərɪə] *n.* *(Psych.)* a neurotic state of uncontrolled excitement or emotion ▶ Mass hysteria was generated in the town by the sound of the approaching bombardment.

hysterics [hɪˈsterɪks] *n. pl.* *(Psych.)* an attack of hysteria ▶ She flew into hysterics at the sight of the crash and had to be sedated.

i

idle [ˈaɪd(ə)l] v.i. (Mech.) (of an engine) to run with the transmission disconnected, so that there is no motion forward or backward ▶ The engine was left to idle for several minutes in order to warm up.

igneous [ˈɪgnɪəs] a. (Geol.) (of rock) produced by volcanic action

ignite [ɪgˈnaɪt] v.t. and v.i. (Phys.) to set alight or to begin to burn ▶ Ignite blue touchpaper and retire. ▶ After contact with the air, the phosphorus ignited and started a fire.

ignition [ɪgˈnɪʃ(ə)n] n. (Mech.) the mechanism in a petrol engine that causes the fuel to ignite ▶ To start the engine you must first switch on the ignition.

ignition key the key that operates the ignition system of a motor engine ▶ To switch on the ignition you must turn the ignition key.

image [ˈɪmɪdʒ] n. (Optics) the figure of an object formed via a mirror or lens by rays of light ▶ What is seen through the lens of a camera is an inverted image.

immerse [ɪˈmɜs] v.t. to plunge something into or under water or another liquid

immersion [ɪˈmɜʃ(ə)n] n. the act or state of being immersed

immersion heater [ɪˈmɜʃ(ə)n ˌhitə] an electrically-heated device placed in a tank in order to heat the water ▶ Many householders prefer to heat water by immersion heaters instead of boilers.

immune [ɪˈmjun] a. (Med.) not liable to catch an infection

immunity [ɪˈmjunɪtɪ] n. (Med.) the state of being immune

immunize [ˈɪmjunaɪz] v.t. (Med.) to make immune to a disease by artificial means, especially inoculation ▶ Children should be immunized against diphtheria at the age of six weeks.

impedance [ɪmˈpid(ə)ns] n. (Elec.) resistance to alternating current

impediment [ɪmˈpedɪmənt] n. (Med.) (of speech) a defect, such as stammering ▶ He was unable to express himself clearly because of a serious speech impediment.

impel [ɪmˈpel] v.t. (Mech.) to drive forward

impervious [ɪmˈpɜvɪəs] a. (of a material, etc.) which cannot be penetrated

impetus [ˈɪmpɪtəs] n. (Phys.) the force with which a body moves or is impelled ▶ The impetus from the following vehicle pushed our car into the one in front.

implant¹ [ɪmpˈlant] v.t. (Med.) to insert something by surgery ▶ The device implanted in the patient's body enabled it to continue to function normally.

implant² [ˈɪmplant] n. (Med.) the organ or device implanted by a surgeon in a body

implode [ɪmˈpləʊd] v.i. (Phys.) to burst inwards, the opposite of explode ▶ The light bulb imploded as it hit the ground.

implosion [ɪmˈpləʊʒ(ə)n] n. (Phys.) the act of imploding

impregnate [ɪmˈpregneɪt] v.t. (Biol.) to make pregnant, fertilize, fill by soaking, etc.

improper fraction (Maths.) a fraction in which the numerator (above the line) is equal to or greater than the denominator (below the line) ▶ An example of an improper fraction is three over two ($\frac{3}{2}$).

impulse [ˈɪmpʌls] n. (Phys.) a temporary force acting in one direction ▶ In a living organism, impulses are transmitted along nerve fibres. ▶ Electrical impulses are sent in series along a wire or through the air.

impure [ɪmˈpjʊə] a. (Chem.) (of a substance) mixed with another substance

impurity [ɪmˈpʊərɪtɪ] *n.* (*Chem.*) a quantity of a substance which is mixed in another, preventing it from being pure ▶ All impurities were removed from the solution before it could be used in the experiment.

inanimate [ɪnˈænɪmət] *a.* (*Biol.*) not living ▶ Trees and animals are animate, but rocks are inanimate.

inboard [ˈɪnbɔd] *a.* (*Mech.*) (of a motor) inside a boat, as opposed to being fixed as an outboard motor on the stern

incandescence [ˌɪnkænˈdes(ə)ns] *n.* (*Phys.*) the state or quality of being incandescent

incandescent [ˌɪnkænˈdes(ə)nt] *a.* (*Phys.: Elec.*) of an electric or other lamp that has a filament or mantle which is made intensely luminous by heat ▶ The incandescent glare almost blinded us.

inch [ɪnʃ] *n.* (*Meas.*) a unit of measurement equivalent to 25·4 mm ▶ Measurement in yards, feet and inches is now being replaced by metres, cm, and mm.

incinerate [ɪnˈsɪnəreɪt] *v.t.* to destroy by fire

incinerator [ɪnˈsɪnəreɪtə] *n.* a receptacle in which rubbish is burned

incise [ɪnˈsaɪz] *v.t.* (*Med.*) to cut into carefully with a sharp instrument, such as a scalpel

incision [ɪnˈsɪʒ(ə)n] *n.* (*Med.*) a sharp or precise cut ▶ The surgeon took a scalpel and made a firm incision in the patient's abdomen.

incisor [ɪnˈsaɪzə] *n.* (*Med.*) a front tooth, used for biting off food

inclination [ˌɪnklɪˈneɪʃ(ə)n] *n.* (*Maths.*) the angle at which two lines or planes meet

inclined plane (*Maths.*) a plane at an angle of less than 90 degrees (90°) to the horizontal

incombustible [ˌɪnkəmˈbʌstəb(ə)l] *a.* (*Phys.*) incapable of being burned ▶ Asbestos is an incombustible material, but it is dangerous to work with.

incompatibility [ˌɪnkəmpætəˈbɪlətɪ] *n.* the state of being incompatible

incompatible [ɪnkəmˈpætəb(ə)l] *a.* (*Med.: Comput.*) unable or unsuitable to function together ▶ The two men's blood groups were incompatible, so a blood transfusion could not be given. ▶ The two computers are incompatible, so their programs cannot be interchanged.

incontinence [ɪnˈkɒntɪnəns] *n.* (*Med.*) the state of being incontinent

incontinent [ɪnˈkɒntɪnənt] *a.* (*Med.*) unable to restrain natural evacuations of the bladder or bowels ▶ Young children, very elderly people, and sufferers from kidney disease, are sometimes incontinent.

incubate [ˈɪŋkjʊbeɪt] *v.t.* and *v.i.* (*Biol.*) (of eggs, embryos, or bacteria) to develop, or cause to develop

incubation [ˌɪŋkjʊˈbeɪʃ(ə)n] *n.* (*Biol.*) the process of incubating ▶ The incubation period of cholera bacteria is only a few hours.

incubator [ˈɪŋkjʊˌbeɪtə] *n.* (*Med.: Mech.*) an apparatus or device in which living organisms are enabled to develop by being kept warmer than the outside environment ▶ The premature baby was placed in an incubator until it had grown to a more normal size and strength.

indefinite [ɪnˈdefɪnət] *a.* (*Maths.*) without definite or conceivable limits

index¹ [ˈɪndeks] *pl.* **indices** [ˈɪndɪsiz] *n.* (*Maths.*) a superscript number which shows how many times a number or letter is to be multiplied by itself ▶ In 4^3 and x^5, the superscript numbers ^{3} and ^{5} are indices ($4^3 = 64$, and where $x = 2$, $x^5 = 32$).

index² *pl.* **indexes** [ˈɪndeksɪz] *n.* (*Meas.*) the number indicating the variation of one thing as opposed to another thing ▶ The varying height of the mercury in a thermometer is an index of the changes in the temperature of the surrounding atmosphere.

indicator [ˈɪndɪkeɪtə] *n.* (*Chem.*) a reagent used to indicate the presence of an acid or alkali by changing colour ▶ Probably the best-known indicator is litmus paper.

indigenous [ɪnˈdɪdʒɪnəs] *a.* (*Biol.*) native to a certain place ▶ Plants indigenous to tropical islands will not flourish in cold climates.

indigestible [ˌɪndɪˈdʒestəbl] *a.* (*Biol.*) incapable of being digested

indigestion [ˌɪndɪˈdʒestʃ(ə)n] *n. (Med.)* difficulty and discomfort in digesting food ▶ Common symptoms of indigestion are abdominal pain and nausea.

indivisible¹ [ˌɪndɪˈvɪzəb(ə)l] *a. (Maths.)* (of a number) which cannot be divided into equal parts without leaving a remainder ▶ 9 is indivisible by 4 (4 + 4 + 1) but not by 3 (3 + 3 + 3).

indivisible² *a. (Phys.)* (of a particle) too infinitely small to be further subdivided

induce [ɪnˈdjus] *v.t. (Med.)* (of labour) to bring on or speed up by artificial means, such as the use of drugs ▶ Because the baby was overdue, the doctors decided to induce labour.

inductance [ɪnˈdʌkt(ə)ns] *n. (Elec.)* the characteristic of an electric circuit by which, as a result of a change of current in the same circuit, or in a nearby one, an electromotive force (EMF) is set up

inductance coil *see* **induction coil**

induction¹ [ɪndˈʌkʃ(ə)n] *n. (Med.)* the bringing on or speeding up of a birth by artificial means, such as the use of drugs

induction² *n. (Elec.)* the production of an electric or magnetic state by the proximity or movement of an electric or magnetized body

induction coil a kind of transformer which produces a high-voltage electric current from low-voltage direct current ▶ Induction coils are used to produce sparks in internal combustion engines.

induction stroke the downward movement of the piston in the cylinder of an internal combustion engine

inert [ɪˈnɜt] *a. (Chem.)* lacking active chemical powers, neutral

inert gas a gas that does not react with other elements to form chemical compounds ▶ The most common inert gases are helium and neon.

inertia [ɪˈnɜʃə] *n. (Phys.)* the tendency of an object to remain where it is, or – if it is in motion – to continue moving in a straight line at the same speed, unless affected by an external force ▶ When a cyclist stops pedalling, the bicycle continues by its own inertia to move forward at the same speed, unless it is slowed or stopped by the brakes, an upward slope, or some other force.

infall [ˈɪnfɔl] *n. (Eng.)* the place at which water enters a reservoir

infantile paralysis *see* **poliomyelitis**

infanticide [ɪnˈfæntɪsaɪd] *n.* the practice of killing new-born children

infantilism [ɪnˈfæntɪlɪz(ə)m] *n. (Psych.)* a state of underdevelopment in which an adult behaves like a child

infarct [ˈɪnfɑkt] *n. (Med.)* bodily tissue which has died because the blood supply has been cut off

infect [ɪnˈfekt] *v.t. (Med.)* to transmit the germs of an illness to someone else

infection [ɪnˈfekʃ(ə)n] *n. (Med.)* transference of a disease from one organism to another via water or atmosphere, or the disease so transferred ▶ She suffered from an upper respiratory infection and through her constant sneezing transmitted it to the rest of the family.

infectious [ɪnˈfekʃəs] *a. (Med.)* (of a disease) likely to spread to others

inferior¹ [ɪnˈfɪərɪə] *a. (Astron.)* having an orbit between the Sun and the Earth; below the horizon

inferior² *a. (Bot.)* growing below another organ, such as the calyx

inferiority complex [ɪnˌfɪərɪˈɒrɪtɪ ˌkɒmpleks] *(Psych.)* a suppressed sense of inferiority, usually shown by some form of abnormal behaviour, such as over-assertiveness ▶ His desire always to be the centre of attention was a sure sign of a hidden inferiority complex.

infest [ɪnˈfest] *v.t.* to swarm over ▶ The thatched roof of the old barn is infested with rats.

infill [ˈɪnfɪl] *v.t.* to build on areas between existing buildings ▶ Every space between the old village cottages was infilled with brash new houses.

infinite [ˈɪnfɪnət] *a. (Maths.)* having to do with a quantity that is too great to be measured

▶ There are an infinite number of tiny organisms in the sea.
infinitesimal [ˌɪnfɪnɪˈtesɪm(ə)l] *a. (Maths.)* too small to measure
inflammable [ɪnˈflæməb(ə)l] *a. (Phys.)* easily ignited ▶ The inflammable material used in the manufacture of the furniture was declared dangerous and its use became illegal (*see* **flammable**).
inflammation [ˌɪnfləˈmeɪʃ(ə)n] *n. (Med.)* an abnormal condition of a part of the body, involving swelling, redness and pain
influenza [ˌɪnfluˈenzə] *n. (Med.) (abbr.* flu) a catarrhal inflammation of the mucous membranes of the air-passages, with fever and nervous prostration ▶ Influenza is highly contagious.
influx [ˈɪnflʌks] *n.* (of water etc.) a flowing in
informatics [ˌɪnfəˈmætɪks] *n. sg. (Comput.)* information science or technology
information *n. (Comput.)* interpreted data
information retrieval storage, classification, and access to computerized information
information science the practice or study of computerized processing and communication of data
information technology the gathering, processing and communication of information through computing and telecommunications combined
information theory mathematical theory concerning the transmission, storage, retrieval and decoding of information
infra- *comb. form* below, beneath
infra-red [ˌɪnfrəˈred] *a. (Phys.)* (of rays) having to do with invisible radiations beyond the visible spectrum, at the red end ▶ Infra-red rays are sometimes used in the treatment of arthritis.
infrasonic [ˌɪnfrəˈsɒnɪk] *a. (Phys.)* having a frequency below the usual audible limit
ingest [ɪnˈdʒest] *v.t. (Med.)* (of food and drink) to take into the body ▶ Because of injuries to her throat, the patient was unable to ingest food normally, so artificial feeding was necessary.
inguinal [ˈɪŋgwɪn(ə)l] *a. (Med.)* referring to the groin ▶ The most common type of rupture is inguinal hernia.
inhalant [ɪnˈheɪl(ə)nt] *n. (Med.)* a vapour which patients inhale for its medicinal effect
inhalation [ˌɪnhəˈleɪʃ(ə)n] *n. (Biol.)* the breathing in of a gas or vapour ▶ Regular inhalation of tobacco smoke damages the lungs.
inhale [ɪnˈheɪl] *v.t.* and *v.i. (Biol.)* to breathe in ▶ It was a joy to inhale the fresh mountain air. ▶ He said that he did occasionally smoke a cigarette, but that he never inhaled.
inhaler [ɪnˈheɪlə] *n. (Med.)* a piece of apparatus that aids inhalation of a medicinal vapour
inhibit [ɪnˈhɪbɪt] *v.t. (Chem.)* to prevent or impede something from happening ▶ The presence of lime in the soil inhibits the growth of plants of the rhododendron family.
inhibitor [ɪnˈhɪbɪtə] *n. (Chem.)* a drug which delays the development of a process ▶ The chemotherapy included a course of inhibitors to combat the advance of the cancer.
inject [ɪnˈdʒekt] *v.t. (Mech.: Med.)* to introduce a fluid into something by mechanical means ▶ Petrol is injected into the chamber and then ignited. ▶ The nurses injected all the inhabitants of the village with an anti-cholera serum.
injection [ɪnˈdʒekʃ(ə)n] *n. (Mech.: Med.)* the introduction of a liquid by means of a jet or a syringe ▶ The fuel injection jet became blocked and the engine died. ▶ The disease was accidentally spread by the use of dirty needles for the injections.
inlet [ˈɪnlet] *n. (Mech.)* an opening through which a fluid can enter a machine or a container ▶ Lubricating oil seeps through the inlets in the sleeve and prevents the moving parts from becoming too hot.
innumerate [ɪˈnjum(ə)rət] *a.* ignorant of or unskilled in mathematics
inoculate [ɪˈnɒkjʊleɪt] *v.t. (Med.)* to inject an animal or human being with a small amount of a vaccine in order to encourage

the build-up of antibodies as protection against catching a particular disease ▶ Babies are inoculated against measles.

inoculation [ɪˌnɒkjʊˈleɪʃ(ə)n] *n.* (*Med.*) the process of inoculating a person or animal ▶ Inoculation against rabies gives protection for about a year.

inoperable [ɪˈnɒp(ə)rəb(ə)l] *a.* (*Med.*) (of an illness) which cannot be operated on surgically ▶ The cancer was found to be inoperable and had to be treated by other means.

inorganic [ˈɪnɔˈgænɪk] *a.* (*Chem.*) (of a compound) not containing carbon, not made of a living substance ▶ Inorganic chemistry deals with non-living things, whereas organic chemistry deals with organisms, plants or animals.

input [ˈɪnpʊt] *n.* (*Elec.: Comput.*) the current or data put into a system

insect [ˈɪnsekt] *n.* (*Biol.*) a small animal, such as a beetle or a fly ▶ Most insects have three pairs of legs, a body in three sections, two antennae, compound eyes, and no backbone.

insecticide [ɪnˈsektɪˌsaɪd] *n.* (*Chem.: Agric.*) a chemical substance used to kill insects considered to be harmful ▶ Over-use of insecticides may be harmful to crops and to other kinds of wildlife.

insectivore [ɪnˈsektɪvɔ] *n.* (*Zool.*) a bird or animal that feeds on insects

insoluble [ɪnˈsɒljʊb(ə)l] *a.* (*Chem.*) (of a substance) that cannot be dissolved ▶ Petroleum products such as oil and tar are insoluble in water.

insomnia [ɪnˈsɒmnɪə] *n.* (*Med.*) inability to sleep at night

insomniac [ɪnˈsɒmnɪæk] *n.* (*Med.*) a person who suffers regularly from insomnia

inspection chamber [ɪnˈspekʃ(ə)n ˌtʃeɪmbə] a shaft through which a person can enter a sewer, big tank or container in order to inspect the inside

inspection pit [ɪnˈspekʃ(ə)n ˌpɪt] a pit over which a motor vehicle may be parked so that its underside may be examined

install [ɪnˈstɔl] *v.t.* (*Eng.: Build.*) to put (apparatus, machinery, buildings etc.) in place ▶ A new computer system has now been installed.

installation [ˈɪnstəˈleɪʃ(ə)n] *n.* (*Eng.: Build.*) the process or result of putting something in place ▶ Despite the public outcry, the installation of a nuclear power station was completed.

insulate[1] [ˈɪnsjʊleɪt] *v.t.* (*Elec.*) to separate something from other things by covering it with a non-conductor such as PVC, so that an electric current will not pass through it ▶ All live electric cables must be properly insulated.

insulate[2] *v.t.* (*Phys.: Build.*) to keep warm or quiet by using a layer of material that resists the penetration of cold or sound ▶ The roof of the house was insulated with a layer of fibreglass, which saved energy by cutting down the heat loss.

insulin [ˈɪnsjʊlɪn] *n.* (*Biol.: Med.*) a substance, produced in the body, which controls the level of sugar in the blood ▶ Diabetes is caused by a deficiency of insulin.

intake [ˈɪnteɪk] *n.* (*Mech.*) a place in a machine where liquid or gas is admitted

integer [ˈɪntɪdʒə] *n.* (*Maths.*) a whole number, as opposed to a fraction ▶ 3 is an integer, but $1\frac{1}{2}$ is not.

integral [ˈɪntɪgr(ə)l] *a.* (*Maths.*) made up of or expressible in terms of integers

integrated circuit (*Elec.: Comput.*) a very small circuit on a chip of semiconductor material

intelligence quotient (abbr. IQ) (*Psych.*) a number, used to denote the ratio of a person's intelligence to the average, obtained by dividing the mental age by the age in years and multiplying by 100

intensity [ɪnˈtensɪtɪ] *n.* (*Elec.*) the amount of current per unit

intensive care [ɪnˌtensɪf ˈkeə] special and continuous care of a very sick patient in hospital ▶ After the operation, the patient spent a week in an intensive care ward.

inter- *comb. form* between, among, with

interactive [ɪntəˈræktɪv] *a.* (*Comput.*) permitting continuous mutual communication between computer and user ▶ The design of interactive programs enables users to develop their skills more effectively.

intercept[1] [ˌɪntəˈsept] v.t. (Maths.) to mark off or include between two points on a line

intercept[2] [ˈɪntəsept] n. (Maths.) the part of a line that is intercepted

intercom [ˈɪntəkɒm] (abbr. of **intercommunication** [ˌɪntəkəˌmjuːnɪˈkeɪʃ(ə)n]) n. (Elec.) a system of short-distance verbal communication within a confined area such as an aircraft

interference [ˌɪntəˈfɪərəns] n. (Radio) the spoiling of radio reception by atmospherics or by other signals ▶ We could hardly make out the message being broadcast because of interference from other stations on almost the same wavelength.

interferon [ˌɪntəˈfɪərɒn] n. (Med.) an anti-viral substance produced in living cells in humans and other creatures in response to viral infections ▶ Medical researchers are attempting to employ synthetic interferon in the treatment of viral diseases.

interlock [ˌɪntəˈlɒk] v.t. (Mech.) to lock or fasten firmly together, so that movement of one part controls also movement of the other part

internal [ɪnˈtɜn(ə)l] a. having to do with the inside

internal combustion engine [ˌɪntɜn(ə)l kəmˈbʌstʃ(ə)n ˌendʒɪn] an engine in which mechanical energy is produced by the combustion or explosion of a mixture of air and petrol vapour in its cylinders ▶ The use of the internal combustion engine in motor vehicles has revolutionized travel.

internal medicine a branch of medicine concerned with internal bodily organs

interruptor [ˌɪntəˈrʌptə] n. (Elec.) a device for breaking and closing an electrical circuit ▶ The insertion of an interruptor in the circuit produced a flashing light for use as a signal.

intersect [ˈɪntəsekt] n. (Maths.) the point at which two lines or planes cut across each other

interval [ˈɪntəv(ə)l] n. (Phys.) the difference of pitch between two sounds

intestine [ɪnˈtestɪn] n. usu. pl. **intestines** (Biol.) the bowels or guts

intra- comb. form within, on the inside

intravascular [ˌɪntrəˈvæskjʊlə] a. (Med.) situated or occurring within a blood vessel ▶ The intra-vascular sediment narrowed the passage in the artery and inhibited the flow of blood.

intravenous [ˌɪntrəˈviːnəs] a. (Med.) into or through a vein ▶ The dehydrated patient was given intravenous fluids.

intro- comb. form into, inward, in

introvert [ˈɪntrəvɜt] n. (Psych.) a person more concerned with their own thoughts than with the external world

intrusion [ɪnˈtruːʒ(ə)n] n. (Geol.) the penetration of volcanic rocks into strata of other origins ▶ A vast volcanic intrusion in the seabed gave rise to a new range of mountains below the surface of the water.

invariable [ɪnˈveərɪəb(ə)l] n. (Maths.) a number or quantity which is constant or remains unchanged

invertebrate [ɪnˈvɜtɪbrət] a. or n. (Biol.) having no backbone ▶ The most common example of an invertebrate is the earthworm.

in vitro [ɪn ˈviːtrəʊ] (Latin) a. (Med.) in an artificial environment outside the body ▶ The practice of *in vitro* fertilization enables hitherto infertile couples to have children.

iodine [ˈaɪədɪn] n. (Chem.: Med.) a non-metallic element formerly used in domestic first-aid medical treatment for its antiseptic and disinfectant qualities, and now used in more sophisticated branches of medicine, in photography, and in the manufacture of dyes

ion [ˈaɪən] n. (Phys.) an electrically charged atom or group of atoms ▶ Ions are produced when a salt is dissolved in water.

ionosphere [aɪˈɒnəsfɪə] n. (Phys.) the region surrounding the earth at a height of from six miles (9·5 km) to about 250 miles (400 km) in which ionized layers of gas occur

IQ *see* **intelligence quotient**

iris [ˈaɪ(ə)rɪs] *n.* (*Biol.*) the circular coloured membrane surrounding the pupil of the eye ▶ What is referred to as the colour of someone's eyes is really the colour of the iris.

iron [ˈaɪən] *n.* (*Chem.: Metall.*) a very tough and malleable metallic element ▶ Iron is used extensively for making tools.

iron foundry [ˈaɪən ˌfaʊndrɪ] a place where iron is smelted and purified (*also called* **ironworks**)

iron lung [ˌaɪən ˈlʌŋ] a device for mechanically assisting or maintaining breathing

ironworks [ˈaɪən ˌwɜːks] *see* **iron foundry**

irradiate [ɪˈreɪdɪeɪt] *v.t.* (*Phys.*) to expose something to specific kinds of rays, such as infra-red ▶ Some packaged foods are irradiated with gamma rays to sterilize and preserve them.

irrigate[1] [ˈɪrɪgeɪt] *v.t.* (*Agric.*) (of land) to water in order to promote the growth of crops ▶ When the neglected fields were irrigated, they became fertile again.

irrigate[2] *v.t.* (*Med.*) to keep a wound moist and free from blood etc. during a surgical operation by the use of a constant flow of antiseptic fluid

irritant [ˈɪrɪtənt] *n.* (*Biol.*) a substance that causes discomfort, such as itching, to the skin or some other part of the body ▶ The insecticide acted as a powerful irritant to the eyes of anyone who came near.

iso- *comb. form* same, equal

isobar [ˈaɪsə(ʊ)ˌbɑː] *n.* (*Meteor.*) a line on a map connecting all places with the same barometric pressure at a given time ▶ The fact that the isobars were very close together indicated that a high wind could be expected in the area.

isochronous [aɪˈsɒkrənəs] *a.* (*Phys.*) occurring at the same time or at equal intervals of time ▶ The swings of a pendulum are roughly isochronous.

isolate[1] [ˈaɪsəleɪt] *v.t.* (*Med.*) to separate or set apart ▶ Patients suffering from infectious diseases are isolated from the others.

isolate[2] *v.t.* (*Chem.*) to obtain in an uncombined form by removing all other substances ▶ The basic element was isolated and then weighed.

isolation [ˌaɪsəˈleɪʃ(ə)n] *a.* and *n.* the process or result of being isolated

isolation hospital a hospital for patients suffering from highly infectious diseases

isolation ward a hospital ward in which patients suffering from infectious diseases are isolated from the other patients

isomer [ˈaɪsəmə] *n.* (*Chem.: Phys.*) one of several compounds whose molecules have the same number of atoms, differently arranged

isomorph [ˈaɪsə(ʊ)mɔːf] *n.* (*Maths.*) one of a number of groups which have identical form and structure

isosceles [aɪˈsɒsəliːz] *a.* (*Maths.*) (of a triangle) having two equal sides

isotherm [ˈaɪsə(ʊ)θɜːm] *n.* (*Phys.*) a line on a map passing over places with the same mean temperature

isotope [ˈaɪsətəʊp] *n.* (*Phys.: Chem.*) one of a set of species of atoms of a chemical element with the same atomic number but a different atomic weight

j

jab [dʒæb] *n.* (*Med.*) *coll.* a vaccination or injection ▶ I had to have a number of jabs when I went on holiday in the tropics.

jack [dʒæk] *n.* (*Mech.*) a device for lifting heavy weights, such as motor vehicles

jack-knife [ˈdʒæknaɪf] *v.i.* (of a lorry or truck with a trailer) to get into a position where the trailer is at such an acute angle to the driver's cab that it cannot be controlled ▶ An articulated lorry had jack-knifed across the carriageway and was causing a lengthy hold-up.

jack up to raise a heavy weight by the use of a jack ▶ He jacked up the rear of the car to change the wheel with a punctured tyre.

jam [dʒæm] *v.t.* (*Radio*) to prevent clear reception of a radio signal by transmitting something on the same wavelength ▶ It was impossible to get a message through because the enemy was jamming the wavelength.

jamb [dʒæm] *n.* (*Build.*) one of the upright sides of a doorway or window ▶ The jamb was so warped by damp that the window would not close properly.

jaundice [ˈdʒɔndɪs] *n.* (*Med.*) a condition in which the skin and eyes become yellow

jaw [dʒɔ] *n.* (*Mech.*) one of the two parts of a tool or machine that can grip or crush an object ▶ He adjusted the jaws of the spanner to grip a larger nut.

JCB [ˌdʒeɪsiˈbi] (*abbr.*) *n.* (*Mech.*) a machine with a hydraulically-operated shovel at the front and an excavator at the back ▶ The JCB is named after its inventor, Joseph Cyril Bamford.

jet[1][dʒet] *n.* (*Mech.*) a stream of water, flame, or gas forced through a small aperture at high pressure ▶ The jets of water from the firemen's hoses were directed on the upstairs windows of the burning building.

jet[2] *n.* (*Mech.*) a narrow opening or aperture through which water, gas etc is forced ▶ The petrol jet was blocked with grit and the car engine would not start.

jet-lag [ˈdʒetˌlæg] *n.* (*Med.*) exhaustion caused by travelling through several time-zones at high speed in a jet plane ▶ He arrived in England from the USA suffering from jet-lag.

jet plane (*abbr.* **jet**) a jet-propelled plane

jet-propelled [ˈdʒetprəˈpeld] (of an aircraft) powered by heating and expanding air, which is expelled through a jet from the rear of the plane

jet-propulsion [ˌdʒetprəˈpʌlʃ(ə)n] the system by which a jet-propelled plane is powered ▶ The invention of jet-propulsion has revolutionized air travel.

jet-stream [ˈdʒet ˌstrim] a stream of very strong winds at a great height ▶ A favourable jet-stream shortened the flight time by almost an hour.

jib [dʒɪb] *n.* (*Mech.*) the extended arm of a crane or derrick ▶ The jibs of the huge dockyard cranes were visible for miles.

jig [dʒɪg] *n.* (*Mech.*) a device for holding an object and guiding a cutting tool, used in the manufacture of standard parts

jigger [ˈdʒɪgə] *n.* (*Mining*) a sieve shaken in water to separate the coal or ore poured into it from waste matter

joist [dʒɔɪst] *n.* (*Build.*) one of a series of parallel horizontal beams, of timber or other material, to which floorboards or the laths of a ceiling are nailed ▶ When the floorboards in the old house were raised, it was discovered that several of the joists were rotten.

jugular [ˈdʒʌgjʊlə] *a.* (*Med.*) having to do with

the neck or throat ▶ The jugular vein was severed and she bled to death.

jump-jet [ˈdʒʌmpˌdʒet] an aircraft that can take off and land vertically ▶ Jump-jets do not need airports with lengthy runways.

jump-start [ˈdʒʌmpˌstɑt] to start the engine of a car etc. by first setting the vehicle in motion and then engaging the gear, or by attaching leads from the battery to that of another car ▶ The battery was flat, so they had to jump-start the car.

junction box [ˈdʒʌnʃ(ə)n ˌbɒks] an earthed box in which wires or cables can be safely connected ▶ Proper use of junction boxes reduces the danger of fire.

k

K, k *(abbr.)* Kelvin scale; kilo-; 1,024 words, bytes or bits

kaolin [ˈkeɪəlɪn] *n. (Geol.: Med.)* fine white clay used in making porcelain and china and as an absorbent in treating disorders of the stomach ▶ His chronic indigestion was relieved by regular doses of a medicine containing kaolin.

karst [kɑst] *n. (Geol.)* a limestone region, with underground streams and caves ▶ The coastline of Dalmatia is a rugged stretch of karst.

kc *(abbr.)* kilocycle

keel [kil] *n. (Naut.)* the bottom part of a ship that extends downwards to improve lateral stability ▶ The ship's keel was damaged by a coral reef when it sailed into shallow water.

keeper [ˈkipə] *n. (Elec.)* a bar of soft iron used to connect the poles of a magnet when not in use, in order to close the circuit and conserve the magnetic power

kelvin [ˈkelvɪn] *a.(Phys)* having to do with a thermometer scale on which zero is absolute zero

keratosis [ˌkerəˈtəʊsɪs] *n. (Med.)* a horny growth on the skin, such as a wart; also the skin condition that leads to such growths

kerosine (kerosene) [ˈkerəˌsin] *n. (Chem.)* a kind of paraffin oil, used mainly in oil lamps

keyboard [ˈkibɔd] *n. (Comput.)* a range of keys, like those of a typewriter, used for setting text which is visible on a screen

keystone [ˈkistəʊn] *n. (Eng.: Build.)* the wedge-shaped stone at the top of an arch that holds the remainder in place

keystroke [ˈkistrəʊk] *n. (Comput.)* the operation of a key on a keyboard-operated machine ▶ Loading a document by computer is costed according to the number of keystrokes.

kg *(abbr.)* kilogram (kilogramme)

kidney [ˈkɪdnɪ] *n. (Biol.: Med.)* one of the two bodily organs that remove waste matter from the body in the form of urine ▶ Failure of the kidneys to function properly can lead to a fatal build-up of toxic waste in the blood.

kidney machine [ˈkɪdnɪ məˌʃin] a piece of medical apparatus which operates in the place of kidneys that are not functioning properly

kidney stone [ˈkɪdnɪ stəʊn] a hard mass in the kidney ▶ The surgeon removed two large kidney stones in the course of the operation.

kiln [kɪln] *n. (Build.)* a furnace for baking or drying bricks, etc.

kilo- *comb. form* 1,000; (of words, bytes, or bits) 1,024

kilobyte [ˈkɪlə(ʊ)baɪt] *n. (Comput.)* a unit of computer storage equal to 1,024 bytes

kilocycle [ˈkɪlə(ʊ)ˌsaɪk(ə)l] *n. (Elec.)* a unit for measuring the frequency of alternating current ▶ 1 kilocycle = 1,000 cycles per second.

kilogram (kilogramme) [ˈkɪlə(ʊ)ˌgræm] *n. (Meas.)* a measure of weight; 1,000 grams

kilohertz [ˈkɪlə(ʊ)ˌhɜts] *n. (Meas.)* a measurement of radio waves; 1,000 hertz

kilolitre [ˈkɪlə(ʊ)ˌlitə] *n. (Meas.)* a measurement of liquid volume; 1,000 litres

kilometre [ˈkɪlə(ʊ)ˌmitə] *n. (Meas.)* a measurement of length or distance; 1,000 metres

kiloton [ˈkɪlə(ʊ)ˌtʌn] *n. (Meas.)* a measurement of weight or power; 1,000 tons

kilovolt [ˈkɪlə(ʊ)ˌvəʊlt] *n. (Meas.)* a unit of electrical force: 1,000 volts

kilowatt [ˈkɪlə(ʊ)ˌwɒt] *n. (Meas.)* a unit of

measurement of electrical energy; 1,000 watts

kinaesthesia (kinesthesia) [ˌkɪnisˈθiziə] *n.* *(Psych.)* the bodily perception of muscular movement *(also called* **kinaesthetics)**

kinaesthetic (kinesthetic) [ˌkɪnisˈθetɪk] *a.* *(Psych.)* having to do with kinaesthesia

kinematics [ˌkɪnəˈmætɪks] *n. sg.* *(Phys.)* the study of movement in space-time without reference to mass or force

kinesi- *comb. form* movement

kinesics [kaɪˈnisɪks] *n. sg.* the study of body movements as non-verbal communication ▶ The use of body language as defined in kinesics can be very revealing when a person is speaking.

kinetic [kaɪˈnetɪk] *a.* *(Phys.)* concerned with movement ▶ Movements such as the flow of water under pressure are said to generate kinetic energy.

kinetics [kaɪˈnetɪks] *n. sg.* *(Phys.)* the study of the movement of bodies in space

kl *(abbr.)* kilolitre

kleptomania [ˌkleptə(ʊ)ˈmeɪnɪə] *n.* *(Psych.)* a form of insanity or mental abnormality that shows in an irresistible compulsion to steal ▶ Though she had plenty of money, she became prone to fits of kleptomania and was several times arrested for shoplifting.

kleptomaniac [ˌkleptə(ʊ)ˈmeɪnɪæk] *a. or n.* *(Psych.)* having to do with kleptomania, or someone suffering from it.

km (abbr.) kilometre

knife switch an electrical switch of the type in which a flat metal blade is pushed between two contacts very close together.

knit [nɪt] *v.i.* *(Med.)* to combine together into a whole ▶ After breaking his leg in a fall, he was confined to bed for several weeks so that the bones could begin to knit.

knocking [ˈnɒkɪŋ] *n.* *(Mech.)* (of an engine) the making of explosive sounds because the mixture in the cylinders has been put under too much pressure ▶ Because of the knocking in the engine, she decided not to drive her car until it had been examined by a mechanic.

knot [nɒt] *n.* *(Meas.)* a unit of speed at sea or in the air, equivalent to one nautical mile, or 1·85 kilometres, per hour ▶ The fastest ship in the fleet was capable of only twenty-five knots.

kph *(abbr.)* kilometres per hour

kryometer *see* **cryometer**

kv *(abbr.)* kilovolt

kW *(abbr.)* kilowatt

kWh *(abbr.)* kilowatt hour

kwashiorkor [ˌkwæʃiˈɔkɔ] *n.* *(Med.)* a nutritional disease caused by lack of protein ▶ Many Ghanaian children suffer from kwashiorkor because of deficiencies in their diet.

l

lab *(abbr.)* laboratory
laboratory [ləˈbɒrət(ə)rɪ] *n. (Chem. etc.)* a building or room in which scientific experiments are conducted or in which chemical articles are manufactured
labour [ˈleɪbə] *n. (Med.)* the process of expulsion of a baby from the uterus during childbirth ▶ She went into labour at about midnight and at 2 a.m. the child was delivered.
labyrinth [ˈlæbərɪnθ] *n. (Med.)* part of the inner ear
lactate [lækˈteɪt] *v.i. (Biol.)* (of a mammal) to produce milk from the breasts
lactic [ˈlæktɪk] *a. (Biol.)* concerned with milk ▶ Lactic acid is used to give flavour to certain drinks.
lactose [ˈlæktəʊs] *n. (Chem.)* the kind of sugar that occurs in milk ▶ Lactose is used in the manufacture of certain baby foods.
lag [læg] *v.t. (Mech.)* to cover pipes, etc., with special materials to prevent heat loss ▶ The water pipes and tank in the roof were lagged to prevent them from freezing in winter.
laminate¹ [ˈlæmɪneɪt] *v.t.* to split, beat or roll materials (e.g. plastic) into thin sheets
laminate² *v.t.* to cover with a thin layer of another substance ▶ The kitchen worksurfaces were laminated with plastic which is easy to keep clean.
lance [lɑns] *v.t. (Med.)* to pierce or cut open ▶ The nurse decided to lance the boil.
lancet [ˈlɑnsɪt] *n. (Med.)* a sharp surgical instrument for cutting ▶ At the sight of the lancet in the nurse's hand, the patient fainted.
landfill [ˈlændfɪl] *n.* the practice of burying rubbish under layers of earth, the rubbish so buried, or the place where it is buried

landlocked [ˈlændlɒkt] *a. (Geog.)* completely surrounded by land, with no direct access to the sea ▶ The Czech Republic is landlocked.
lanyard [ˈlænjəd] *n. (Naut.)* a short cord or line for lashing things together
laparoscope [ˈlæpərəskəʊp] *n. (Med.)* an optical instrument for examining internal parts of the abdomen
larva [ˈlɑvə] *pl.* **larvae** [ˈlɑvi] *n. (Biol.)* the first state of an insect when it issues from an egg ▶ Grubs, caterpillars and maggots are all larvae.
laryngitis [ˌlærɪnˈdʒaɪtɪs] *n. (Med.)* inflammation of the larynx ▶ The guest speaker was suffering from laryngitis and was too hoarse to deliver his speech.
laryngoscope [laˈrɪŋgəskəʊp] *n. (Med.)* a piece of apparatus for examination of the larynx
larynx [ˈlærɪŋks] *n. (Med.)* the upper part of the windpipe. containing the vocal cords
laser [ˈleɪzə] *n. (Phys.)* a powerful and intense beam of light ▶ Lasers have a number of applications, including surgery and missile technology.
laser printer a device for rapid and clear printing of computer documents
latent [ˈleɪt(ə)nt] *a.* hidden or concealed
latent heat heat absorbed or emitted when a liquid turns into a solid, or vice versa
latex [ˈleɪteks] *n. (Biol.: Chem.)* a milky substance, obtained from certain plants and trees ▶ Latex is used in making glue, paints, rubber, etc.
lathe [leɪð] *n. (Mech.)* a machine for shaping metal or wood by revolving it against a cutting surface
latitude [ˈlætɪtjud] *n. (Geog.)* distance measured in degrees north or south of the equator ▶ The border between the USA and

Canada is at a latitude of 49 degrees north (49°N).

launch [lɔntʃ] *v.t.* *(Aer.)* to send a rocket with a spacecraft or missile into space ▶ The space shuttle was launched from Cape Canaveral, in Florida.

launching pad [ˈlɔntʃɪŋ ˌpæd] the structure or platform from which a rocket is launched

lava [ˈlɑvə] *n.* *(Geol.)* melted rock thrown up by a volcano ▶ The molten lava set fire to everything in its path.

laxative [ˈlæksətɪv] *n.* or *a.* *(Med.)* (of a medicine) which facilitates opening the bowels

layer [ˈleɪə] *v.t.* *(Hort.)* to propagate a plant by laying a shoot along the ground, just below the surface

lb *(abbr.)* pound (weight)

LCM [ˈelsiˈem] *(abbr.)* *n.* *(Maths.)* lowest common multiple

leach [litʃ] *v.t.* *(Chem.)* to separate or wash out a soluble substance from another by percolating water through it ▶ The heavy rain leached the nitrates out of the soil.

lead¹ [lid] *n.* *(Elec.)* a wire or cable used to connect electrical appliances to each other or to the power source

lead² [led] *n.* *(Chem.: Metall.)* a soft, heavy and malleable metallic element ▶ In most of its former uses, lead has now been replaced by plastic.

lead-free (of petrol) unleaded, without an admixture of lead ▶ Lead-free petrol can now be used with most cars and is more environment-friendly.

lens [lenz] *n.* *(Phys.)* a piece of glass or another substance with one or both surfaces curved so as to bend the rays of light going through it ▶ When spectacles are made, the lenses are specially ground to suit the eyes of the wearer.

leprosy [ˈleprəsɪ] *n.* *(Med.)* a highly contagious skin disease characterized by ulcers and loss of feeling ▶ Because of advances in medical science, leprosy is now confined to only a few areas of the world.

lesbian [ˈlezbɪən] *n.* a female homosexual

lesion [ˈliʒ(ə)n] *n.* *(Med.)* physical change in an organ or tissue due to injury ▶ There were a number of herpetic lesions on the child's lips.

leukaemia [ljuˈkimɪə] *n.* *(Med.)* a cancer of the blood, in which there is uncontrolled multiplication of the white cells ▶ Some people believe that living close to nuclear power plants increases the danger of contracting leukaemia.

lever [livə] *n.* *(Mech.)* a bar of wood or metal with a fixed fulcrum, for lifting a heavy object by inserting one end under it and pressing downward on the other end

life [laɪf] *n.* *(Biol.)* the state in which an organism is capable of performing its functions

life-cycle [ˈlaɪfˌsaɪk(ə)l] the stages through which a living organism passes ▶ The chrysalis is only one stage in the life-cycle of a butterfly.

life-expectancy [ˈlaɪfɪkˌspektənsɪ] the length of time through which an organism can reasonably be expected to remain alive ▶ The life-expectancy of people living in developed societies has risen steadily throughout the twentieth century.

life science [ˈlaɪf saɪəns] any branch of science, such as Botany and Zoology concerned with the study of living organisms

lifespan [ˈlaɪfˌspæn] *n.* *(Biol.)* the period of time through which an organism normally lives ▶ The lifespan of the elephant is twice as long as that of a human being.

lift [lɪft] *n.* *(Aer.)* the lifting effect of a flow of air on the wing of an aircraft

lift-off the successful launching of a rocket into space

ligament [ˈlɪgəmənt] *n.* *(Biol.)* a short band of fibrous tissue by which the bones are held together ▶ A torn ligament is a common but painful injury suffered by many athletes.

ligature [ˈlɪgətʃə] *n.* *(Med.)* a thread or cord used in surgery to tie blood vessels

light year [ˈlaɪt ˌjɪə] the distance travelled by light in one year ▶ The speed of light is calculated as 300,000 km per second.

lighter [ˈlaɪtə] *n.* *(Naut.)* a flat-bottomed boat used for ferrying cargo to and from a ship

lightning [ˈlaɪtnɪŋ] *n. (Phys.: Meteor.)* a dazzling flash of light caused by an electrical discharge between clouds or from cloud to earth, usually accompanied by thunder ▶ Sheet lightning moves from cloud to cloud, and forked lightning from cloud to earth.

lightning conductor [ˈlaɪtnɪŋ kənˌdʌktə] a metal rod or lead which protects buildings by acting as an earth to the electrical discharge of lightning

lignite [ˈlɪɡnaɪt] *n. (Geol.)* a kind of soft, brown coal, nearer to wood than other kinds

lime [laɪm] *n. (Chem.)* a kind of earth containing a high proportion of calcium oxide ▶ In various forms, lime is used as a fertilizer in agriculture, as a caustic, in making cement for building, and in whitewash for painting farm buildings.

lime-kiln [ˈlaɪm ˌkɪln] a kind of large oven in which limestone is reduced to lime

limestone [ˈlaɪmˌstəʊn] *n. (Geol.)* a type of sedimentary rock consisting mainly of calcium carbonate ▶ Fossils of plants, crustaceans, and even dinosaurs are to be found in limestone.

linctus [ˈlɪŋktəs] *n. (Med.)* a syrupy, sedative cough medicine

linear [ˈlɪnɪə] *a. (Maths.)* (of functions) able to be represented on a graph by a straight line

liniment [ˈlɪnɪmənt] *n. (Med.)* an oily liquid used for rubbing on sore muscles

lint [lɪnt] *n. (Med.)* a soft, absorbent material for covering and protecting cuts and wounds

lintel [ˈlɪnt(ə)l] *n. (Build.)* a horizontal, load-bearing beam above a door or window

liquefy [ˈlɪkwɪfaɪ] *v.t.* and *v.i. (Chem.)* to turn from a solid to a liquid ▶ The tar on the road surface liquefied in the hot sun.

liquid [ˈlɪkwɪd] *n.* and *a. (Phys.)* a substance other than a gas that is fluid or capable of flowing, like water

lithosphere [ˈlɪθə(ʊ)sfɪə] *n. (Geol.)* the hard outer layer of Earth, consisting of rocks and soil

litmus [ˈlɪtməs] *n. (Chem.)* a powdery substance which turns red when it is in contact with an acid and blue when in contact with an alkali

litmus paper [ˈlɪtməs ˌpeɪpə] paper impregnated with litmus and used to indicate the presence of an acid or an alkali

litre [ˈliːtə] *n. (Meas.)* a metric unit of capacity of a liquid (about 1·75 pints)

littoral [ˈlɪtər(ə)l] *n.* and *a. (Geog.)* the shore of the sea or of a large lake

live [laɪv] *a. (Elec.)* (of a wire) carrying an electric charge; (of a bomb) with a detonator

liver [ˈlɪvə] *n. (Biol.)* an internal bodily organ which secretes bile and purifies the blood

load^1 [ləʊd] *v.t. (Comput.)* to transfer a program from a disk, etc., to the computer memory

load^2 *n. (Elec.: Mech.)* the power output of a machine or circuit at a given time

load-bearing (of a beam, girder, or wall) on which structural weight rests

locomotion [ˌləʊkəˈməʊʃ(ə)n] *n. (Phys.)* the act or power of moving from place to place

locomotive^1 [ˌləʊkəˈməʊtɪv] *a. (Phys.)* having the power of movement or causing movement

locomotive^2 *n. (Eng.)* a railway engine or similar machine

log *(abbr.)* **logarithm**

logarithm [ˈlɒɡərɪð(ə)m] *n. (Maths.)* a tabulated exponent in a system of simplifying arithmetical processes by making it possible to substitute addition and subtraction for multiplication and division

long-sighted [ˌlɒŋˈsaɪtɪd] (of a person) able to see to a great distance

longitude [ˈlɒn(d)ʒɪtjuːd] *n. (Geog.)* distance measured in degrees east or west of an imaginary line joining the two poles and passing through Greenwich [ˈɡrenɪtʃ], England

loom [luːm] *n. (Mech.)* a machine for weaving cloth

loop [luːp] *n. (Comput.)* a set of instructions repeated in a program until the intended result is achieved

lowest common denominator the smallest

number that is exactly divisible by every denominator in a series of fractions ▶ The lowest common denominator of $\frac{1}{4}$ and $\frac{1}{3}$ is 12.

lowest common multiple the lowest number that contains two or more numbers an exact number of times without a remainder ▶ The lowest common multiple of 3 and 4 is 12.

LP [ˌelˈpi] *(abbr.)* long playing gramophone record or disc

LSD [ˌelesˈdi] *(abbr.) n. (Chem.: Med.)* lysergic acid diethylamide, a hallucinogenic drug

lubricant [ˈlubrɪkənt] *n. (Mech.)* a substance such as oil or grease which is used to lubricate the moving parts of a machine

lubricate [ˈlubrɪkeɪt] *v.t. (Mech.)* to make the moving parts of a machine smooth and slippery to avoid friction and heat

lumbago [lʌmˈbeɪgəʊ] *n. (Med.)* pain in the lower back

luminosity [ˌlumɪˈnɒsətɪ] *n. (Astron.)* a measure of the amount of light emitted by a star

lunar [ˈlunə] *a. (Astron.)* having to do with the moon ▶ The lengths of the lunar months are different from those of the months in the solar calendar.

lunar module a part of a spacecraft which detaches itself in order to land on the moon

lung [lʌŋ] *n. (Med.)* one of a pair of organs in the chest of vertebrate animals by which they breathe

lymph [lɪmf] *n. (Biol.)* a colourless fluid present in the tissues and organs of the body

lymph gland [ˈlɪmf ˈglænd] one of the glands that secrete lymph

lysergic acid diethylamide *see* **LSD**

m

macadam [məˈkædəm] *n. (Eng.)* a type of road surface made up of small stones bound together with tar or asphalt ▶ Macadam is called after its inventor, John McAdam.

machine [məˈʃin] *n. (Mech.)* a mechanical apparatus used in performing actions that require motive force ▶ The use of machines enables us to do things more quickly and more efficiently than we could do them by hand.

machine-readable (of data) presented in a form usable by a computer

machine-tool [məˈʃinˌtul] a machine, driven by electrical or other power, which is used to perform work otherwise done with a manual tool

machinery [məˈʃinərɪ] *n. (Mech.)* parts or numbers of machines

macro- *comb. form* big, large

macrocosm [ˈmækrə(ʊ)ˌkɒz(ə)m] *n.* the whole universe, with all the parts taken together

magazine[1] [ˌmægəˈzin] *n. (Mech.)* the container for bullets, shells etc. in a light gun

magazine[2] *n.* a building for storing weapons and explosives

maggot [ˈmægət] *n. (Biol.)* the worm-like larva of an insect ▶ Maggots are used as live bait by anglers hoping to catch certain types of fish.

magnesium [mægˈniziəm] *n. (Chem.)* a white metallic chemical element that burns with a brilliant white light ▶ Magnesium was used in early photography to make a bright light.

magnet [ˈmægnɪt] *n. (Phys.)* an object, usually of iron or steel, which has the power of attracting other metal objects

magnetic [mægˈnetɪk] *a. (Phys.)* having the attracting qualities of a magnet

magnetic field the area around a magnet in which its magnetic force is effective

magnetic needle the magnetized needle of a compass, which always points to magnetic north and south

magnetic pole one of the two points on the Earth's surface to which a magnetic needle points ▶ The magnetic poles are not exactly the true north and south poles.

magnetize [ˈmægnɪtaɪz] *v.t. (Phys.)* to make an object magnetic

magneto [mægˈnitəʊ] *n. (Elec.)* a small generator for producing electric sparks from a magnetic field ▶ The magneto is what produces the sparks to ignite the mixture in a petrol engine.

magnify [ˈmægnɪfaɪ] *v.t. (Phys.: Optics)* to make something seem larger than it is by viewing it through a lens ▶ Minute organisms cam be magnified hundreds of times and examined under a microscope.

magnitude [ˈmægnɪtjud] *n. (Maths.)* comparative size ▶ The magnitude of distant stars can now be measured by the use of radio telescopes.

mainframe [ˈmeɪnˌfreɪm] *a.* or *n. (Comput.)* (of a computer) the central storage and processing installation of a computer system ▶ All the computer terminals throughout the campus were linked to the mainframe in the computer centre.

mains [meɪnz] *n. pl. (Eng.)* the main conduits by which electricity, gas, or water are supplied to a building ▶ The old farmhouse was not yet connected to the mains for its water.

mal- *comb. form* badly, wrongly

malaria [məˈleərɪə] *n. (Med.)* a dangerous infectious disease carried by certain types of mosquito ▶ Malaria is characterized by recurrent attacks of fever.

malformation [ˌmælfɔˈmeɪʃ(ə)n] *n. (Med.)* a badly or wrongly formed part of the body ▶ The new-born baby had a malformation of the left leg.

malfunction [ˌmælˈfʌŋʃ(ə)n] *n. (Mech.)* failure or breakdown of a mechanical process ▶ Because of a malfunction of the computer, everyone was paid too much.

malignant [məˈlɪgnənt] *a. (Med.)* (of a disease, tumour, etc.) life-threatening ▶ Too much exposure to sunshine may cause malignant growths on the skin.

malleable [ˈmælɪəb(ə)l] *a. (Metall.)* (of metals) capable of being worked in various ways ▶ Malleable metals can be pressed, rolled or beaten into the desired shapes.

malnutrition [ˌmælnjuˈtrɪʃ(ə)n] *n. (Med.)* an unhealthy condition of ill-health resulting from shortage of the right kind of food ▶ Millions of people throughout the world are suffering from malnutrition, while many of the remainder eat too much for their own welfare.

mammal [ˈmæm(ə)l] *n. (Biol.)* an animal, the female of which gives birth to live young and feeds them with milk from her breasts

manganese [ˈmæŋgəniz] *n. (Chem.)* a hard chemical element used in making steel and glass

manhole [ˈmænhəʊl] *n. (Eng.)* a hole in a floor, street surface etc. through which one may enter a sewer etc. in order to inspect or repair ▶ The cover of the manhole had not been properly replaced and was a danger to passing traffic.

mania [ˈmeɪnɪə] *n. (Psych.)* a form of extreme mental disorder, frequently leading to violence

maniac [ˈmeɪnɪæk] *n. (Psych.)* a person suffering from a mania

manic [ˈmænɪk] *a. (Psych.)* concerning mania

manic depressive [ˌmænɪk dɪˈpresɪv] a person suffering from a mental disorder in which extreme confidence alternates with deep depression

manifest [ˈmænɪfest] *n.* a list of passengers and goods carried on a ship or plane

manifold [ˈmænɪfəʊld] *n. (Mech.)* a pipe or container having various outlets connected with different parts of an engine ▶ Gases escape through outlets in the exhaust manifold.

manipulate [məˈnɪpjʊleɪt] *v.t. (Mech.)* to operate (a piece of apparatus) successfully

manometer [məˈnɒmɪtə] *n. (Phys.)* a U-shaped tube for measuring the pressure of liquids and gases

mantle [ˈmænt(ə)l] *n. (Geol.)* the part of the Earth that surrounds the core beneath the crust

manual [ˈmænjʊəl] *n.* a book of technical instructions ▶ Every new car is supplied with a manual explaining the specifications etc. of the engine.

marine [məˈrin] *a. (Biol.)* having to do with the sea ▶ Biologists are still discovering new kinds of marine creatures and vegetation.

maritime [ˈmærɪtaɪm] *a. (Naut.)* having to do with ships and navigation ▶ A conference of maritime nations met to consider questions of maritime law.

marl [mɑl] *n. (Geol.)* clay containing a high lime content, used as a fertilizer

marrow [ˈmærəʊ] *n. (Med.)* a fatty substance found inside bones ▶ A malignant form of cancer attacks the marrow and is curable only by transplant.

marshal [ˈmɑʃ(ə)l] *v.t.* to assemble (e.g. railway trucks) in the required order

marshalling yard [ˈmɑʃ(ə)lɪŋ ˌjɑd] a place where railway trucks are sorted into the required trains

marsupial [mɑˈsupɪəl] *n. (Zool.)* a type of mammal, the female of which carries the young in a pouch ▶ Among the best known marsupials are kangaroos and wallabies.

masochism [ˈmæsə(ʊ)ˌkɪz(ə)m] *n. (Psych.)* a mental condition in which sexual pleasure is derived from experiencing physical pain

mass [mæs] *n. (Phys.)* the amount of matter that a body contains ▶ Mass is defined in terms of relative speed.

massif [ˈmæsif] *n. (Geol.: Geog.)* the main or central mass of a range of mountains ▶

The mountain range in central France is often referred to simply as the *massif*.

mastectomy [mæˈstektəmɪ] *n. (Med.)* surgical removal of a breast

mathematical [ˌmæθəˈmætɪk(ə)l] *a. (Maths.)* having to do with mathematics ▶ A series of complicated mathematical calculations was necessary to find the solution to the problem.

mathematician [ˌmæθ(ə)məˈtɪʃ(ə)n] *n. (Maths.)* a person with special skill in or knowledge of mathematics

mathematics [ˌmæθ(ə)ˈmætɪks] *n. (Maths.)* the systematic study of numbers, quantities, shapes etc. and their relationships ▶ The principal branches of mathematics are algebra, arithmetic, geometry, trigonometry, and calculus.

matrix¹ [ˈmeɪtrɪks] *pl.* **matrices** [ˈmeɪtrɪsiz] *n. (Maths.)* an arrangement, usually in the shape of a rectangle, of numbers or algebraic symbols related to one another

matrix² *n. (Eng.)* a mould into which liquid metal, vinyl, or other substance is poured in order to produce an article with a specific shape ▶ Matrices are used in the production of gramophone records (discs).

matrix printer a system of printing, in which characters are formed by dots when a pin strikes an inked ribbon on to paper ▶ Many word-processors have matrix printers, which are slow but reliable.

matt [mæt] *a.* dull in appearance, not glossy

matter [ˈmætə] *n. (Phys.)* anything that has weight and mass, occupies space, and can be seen

maxi- *comb. form* very large or very long

maximum [ˈmæksɪməm] *a.* or *n. (Maths.)* the greatest possible quantity or degree that can be obtained in a given case

mean [min] *a.* or *n. (Maths.)* a quantity in between two extremes, an average ▶ The mean midday temperature on the island is almost 30°C.

mechanic [məˈkænɪk] *n. (Mech.)* someone skilled in the maintenance and repair of machinery

mechanical [məˈkænɪk(ə)l] *a. (Mech.)* having to do with or controlled by machinery ▶ The foundations were excavated with a mechanical shovel.

mechanical engineering the branch of engineering having to do with the design and production of machines

mechanism [ˈmekənɪz(ə)m] *n. (Mech.: Biol.)* a system of parts working together, as in a machine

mechanize [ˈmek(ə)naɪz] *v.t. (Mech.)* to convert the performance of activities from manual operation to the use of machines ▶ The entire manufacturing process at the factory has since been mechanized.

median [ˈmidɪən] *n. (Maths.)* a straight line joining the apex of a triangle to the middle point of the opposite side

medical [ˈmedɪk(ə)l] *a. (Med.)* having to do with medicine

medication [ˌmedɪˈkeɪʃ(ə)n] *n. (Med.)* a medicine or course of treatment ▶ When the patient failed to respond, it was decided to change the medication.

medicinal [məˈdɪsɪn(ə)l] *a. (Med.)* beneficial to health ▶ Some mineral waters are thought to have medicinal qualities.

medicine¹ [ˈmed(ɪ)s(ə)n] *n.* the science of preserving health and curing or alleviating disease ▶ Enormous advances in medicine have been achieved in the present century.

medicine² *n. (Med.)* a substance prescribed by a doctor to alleviate or cure a disease ▶ She was advised to take one dose of the medicine three times a day.

medium [ˈmidɪəm] *n. (Phys.)* a substance or the surroundings in which things live, or through which a force is transmitted ▶ Sound travels less well through the medium of water than through the medium of air.

mega-, megalo- *comb. form* very big; one million

megabit [ˈmegəbɪt] *n. (Comput.)* a million bits (2^{20})

megabyte [ˈmegəbaɪt] *n. (Comput.)* a million bytes (2^{20})

megahertz [ˈmegəˌhɜts] *n. (Radio)* a unit of

megalomania [ˌmegələ(ʊ)ˈmeɪnɪə] *n.* (*Psych.*) a mental disorder in which people greatly exaggerate their own importance or power

membrane [ˈmembreɪn] *n.* (*Biol.*) a thin layer of fibrous tissue covering or connecting parts of an animal or plant

memory [ˈmemərɪ] *n.* (*Comput.*) a device for storing data which can be recalled ▶ Each new generation of computers has a greater capacity for storage in its memory.

Mendelism [ˈmend(ə)lɪz(ə)m] *n.* (*Biol.*) a theory of heredity propounded by the Austrian botanist G.J. Mendel (1822–84).

meningitis [ˌmenɪnˈdʒaɪtɪs] *n.* (*Med.*) inflammation of one of the membranes enclosing the brain and the spinal cord

menopause [ˈmenə(ʊ)pɔːz] *n.* (*Biol.: Med.*) the period in a female's life when menstruation and the capacity for reproduction cease ▶ A popular term for the menopause is 'change of life'.

menstruation [ˌmenstruˈeɪʃ(ə)n] *n.* (*Biol.: Med.*) periodic flow of blood from the uterus

mental [ˈment(ə)l] *a.* (*Psych.*) having to do with the mind

mental age the level of a person's mental ability, expressed in terms of the average age of other people with the same mental ability ▶ By his fourteenth birthday he had still only reached the mental age of five.

mercury [ˈmɜːkjʊrɪ] *n.* (*Chem.*) a liquid, silvery-white, poisonous metallic chemical element ▶ Mercury is used in thermometers and barometers.

meridian¹ [məˈrɪdɪən] *n.* (*Geog.*) any imaginary line around the Earth's surface which passes through both the north and the south poles ▶ Longitude is measured by reference to the Greenwich meridian – the imaginary line that passes through Greenwich [ˈgrenɪtʃ], England.

meridian² (*Astron.*) the apparently highest point in the visible sky reached by the Sun or some other heavenly body, or when they appear to cross a given meridian

mesh [meʃ] *v.i.* (*Mech.*) (of cogs) to engage with other cogs in a machine

mesosphere [ˈmezə(ʊ)ˌsfɪə] *n.* (*Phys.*) the region of the Earth's atmosphere above the stratosphere

meta- *comb. form* with, after, between, among

metabolic [metəˈbɒlɪk] *a.* (*Biol.: Chem.*) having to do with metabolism

metabolism [məˈtæbəlɪzm] *n.* (*Biol.: Chem.*) the chemical processes that enable an organism to use energy and function properly ▶ The rate of metabolism may be affected by age and health.

metal [ˈmetl] *n.* (*Chem.*) a mineral substance, such as iron, copper, silver or gold, which conducts electricity and heat and can be shaped in various ways

metallic [məˈtælɪk] *a.* (*Chem.*) having to do with or made of metal

metallurgical [ˌmetəˈlɜːdʒɪk(ə)l] *a.* (*Metall.*) having to do with metallurgy

metallurgy [ˈmet(ə)lɜːdʒɪ] *n.* the scientific study of the extraction of metals from ores and of ways of using them in manufacture

metamorphosis [ˌmetəˈmɔːfəsɪs] *n.* (*Biol.*) a very great and rapid change of form ▶ The metamorphosis of a larva into an insect involves the destruction of old tissue and the growth of new.

metastasis [məˈtæstəsɪs] *n.* (*Med.*) (of cells, especially cancer cells) the spreading from one part of the body to other parts

metastasize [məˈtæstəsaɪz] *v.i.* (*Med.*) (of cells) to spread from one part of the body to another

meteor [ˈmiːtɪə] *n.* (*Geol.: Astron.*) a relatively small rock or piece of metal entering the Earth's atmosphere from outer space and burning up, giving off a bright light ▶ Meteors are known to most people as shooting stars.

meteorite [ˈmiːtɪəraɪt] *n.* (*Astron.: Geol.*) a meteor that reaches earth ▶ Meteorites have caused huge craters where they landed.

meteorology [ˌmiːtɪəˈrɒlədʒɪ] *n.* the scientific study of the atmosphere and of weather and climate ▶ An obvious application of meteorology is forecasting weather.

meter ['miːtə] *n. (Mech.)* an instrument for measuring ▶ Meters are in everyday household use to measure gas consumption, electrical current, wind strength, etc.

methane ['miːθeɪn] *n. (Chem.)* a colourless and odourless gas which burns very readily ▶ In coal mines the presence of methane is a constant danger.

metre ['miːtə] *n. (Meas.)* a unit of length in the metric system ▶ A metre equals 100 cms or 1,000 mm (1·094 yards).

metric ['metrɪk] *a. (Meas.)* having to do with the system using the metre as a basic unit

metric ton 1,000 kilograms

metrication [ˌmetrɪ'keɪʃ(ə)n] *n. (Meas.)* the conversion of a system of measurement to the metric system

metric system ['metrɪk ˌsɪstəm] a system of weights and measures based on divisions of 10 ▶ In the metric system, the basic units are grams (weight), metres (length), and litres (volume).

mg *(abbr.)* milligram

mica ['maɪkə] *n. (Chem.)* one of a group of silicates which can be split into thin sheets ▶ Mica is transparent and was formerly used for making windows.

micro *(abbr.)* microcomputer

micro- *comb. form* very small; one-millionth

microbe ['maɪkrəʊb] *n. (Biol.)* an extremely small organism, visible only through a microscope ▶ The most commonly known microbes are bacteria.

microbiology [ˌmaɪkrə(ʊ)baɪ'ɒləʤɪ] *n. (Biol.)* the scientific study of extremely small organisms

microchip ['maɪkrə(ʊ)ˌtʃɪp] *n. (Comput.)* a chip, usually made of silicon, carrying a large number of circuits

microcircuit ['maɪkrə(ʊ)ˌsəːkɪt] *n. (Elec.)* a very small circuit on a semiconductor

microcomputer ['maɪkrə(ʊ)kəmˌpjuːtə] *n. (Comput.)* a small computer with one or more microprocessors

microdot ['maɪkrə(ʊ)ˌdɒt] *n. (Photo.)* a photographic image reduced to the size of a dot ▶ Microdots are very popular in spy stories.

microelectronics [ˌmaɪkrə(ʊ)ɪlek'trɒnɪks] *n. sg. (Elec.)* the electronics of microcircuits

microfiche ['maɪkrə(ʊ)ˌfiːʃ] *n. (Photo.)* a film with very small reductions of a number of documents ▶ Documents on microfiche can be read on a machine called a microfiche reader.

micrometer [maɪ'krɒmɪtə] *n. (Mech.)* an instrument for measuring very small objects and distances

micro-organism [ˌmaɪkrəʊ'ɔːgəniz(ə)m] *n. (Biol.)* an organism of microscopic size

microphone ['maɪkrəfəʊn] *n. (Phys.: Mech.)* a device for converting sound into electrical waves, which can then be amplified, recorded or broadcast

microprocessor ['maɪkrə(ʊ)ˌprəʊsesə] *n. (Comput.)* an integrated circuit which forms the central processing unit of a microcomputer

microscope ['maɪkrəskəʊp] *n. (Phys.: Mech.)* a piece of apparatus which magnifies small objects for examination, etc.

microscopic [ˌmaɪkrə'skɒpɪk] *a. (Phys.)* so small that it can be seen only through a microscope

microsecond ['maɪkrə(ʊ)ˌsek(ə)nd] *n. (Phys.)* one-millionth of a second

microsurgery [ˌmaɪkrə(ʊ)'səːʤərɪ] *n. (Med.)* surgery performed by using very small instruments and a form of microscope

microwave ['maɪkrə(ʊ)ˌweɪv] *n. (Phys.)* an electromagnetic wave with a wavelength of between 30cm and 1mm ▶ A fast method of cooking certain dishes is the microwave oven, often called simply microwave.

micturate ['mɪktʃʊreɪt] *v.i. (Med.)* to urinate

migraine ['miːgreɪn] *n. (Med.)* a severe headache, usually affecting one side of the head more than the other, often with nausea and disturbances in vision ▶ For many years she suffered from recurrent migraine, for which there was no obvious cause.

migrate [maɪ'greɪt] *v.i. (Biol.)* (of birds and animals) to move from one habitat to another at certain times of the year ▶

Every autumn millions of birds migrate from northern Europe to warmer countries in the South.

migratory [maɪˈgreɪtərɪ] *a.* (*Biol.*) (of birds and animals) that make annual migrations

mile [maɪl] *n.* (*Meas.*) a measure of length equivalent to 1607 metres.

mileometer [maɪlˈɒmɪtə] *n.* (*Mech.*) a device that measures the number of miles a vehicle has travelled

milli- *comb. form* one-thousandth

millibar [ˈmɪlɪbɑ] *n.* (*Phys.*) a unit of atmospheric pressure, one-thousandth of a bar, which is the height of a column of mercury of about 0·762 mm (0·03 inches).

milligram (milligramme) [ˈmɪlɪgræm] *n.* (*Meas.*) one-thousandth of a gram

millilitre [ˈmɪlɪˌlitə] *n.* (*Meas.*) one-thousandth of a litre

millimetre [ˈmɪlɪˌmitə] *n.* (*Meas.*) one thousandth of a metre

mine¹ [maɪn] *n.* (*Mining*) a shaft or drift in the earth from which ores, coal etc. are dug

mine² *n.* (*Chem.: Mech.*) an explosive device, usually concealed just below the surface of the sea or the ground, which explodes on contact

minefield [ˈmaɪnfild] *n.* an area of sea or land in which mines have been placed

mineral [ˈmɪnər(ə)l] *n.* (*Geol.*) any substance, occurring naturally, which is neither a plant (vegetable) nor an animal ▶ Vast deposits of minerals are known to exist in Antarctica, but they would be extremely difficult to mine.

mineral water [ˈmɪn(ə)r(ə)l ˌwɔtə] water naturally impregnated with minerals ▶ Drinking mineral waters is supposed by many to be beneficial to health.

mineralogy [ˌmɪnəˈrælədʒɪ] *n.* (*Geol.*) the scientific study of minerals

mini- *comb. form* small, smaller

minicomputer [ˌmɪnɪkəmˈpjutə] *n.* (*Comput.*) a small digital computer

minimum [ˈmɪnɪməm] *n. or a.* (*Maths.*) the smallest possible quantity or degree that can be obtained in a given case

minus¹ [ˈmaɪnəs] *prep. or a.* (*Maths.*) reduced by... ▶ Twelve minus three equals nine (12−3=9).

minus² *prep. or a.* (*Phys.*) below zero ▶ The temperature in mid-winter fell to minus thirty (−30°C).

minute¹ [ˈmɪnɪt] *n.* (*Meas.*) a unit of time, equivalent to 60 seconds ▶ There are 60 seconds in a minute and 60 minutes in an hour.

minute² *n.* (*Maths.*) (in measuring angles) the sixtieth part of a degree ▶ There are 90 degrees (90°) in a right angle and 60 minutes (60′) in a degree.

missile [ˈmɪsaɪl] *n.* (*Phys.: Mech.*) (usually) a rocket-propelled warhead

mixture¹ [ˈmɪkstʃə] *n.* (*Chem.*) a substance consisting of two other substances which, though they are difficult to separate, have not combined chemically to form a compound

mixture² (*Mech.*) the mixture of petrol fumes and air which is ignited to provide the explosive force in an internal combustion engine ▶ If there is too little fuel in the mixture, the engine will not start.

mm. (*abbr.*) millimetre(s)

model [ˈmɒdl] *n.* a theory or set of assumptions intended to predict future developments

modification [ˌmɒdɪfɪˈkeɪʃ(ə)n] *n.* a small-scale change ▶ Study of the computer model led to several modifications in the design of the project.

modify [ˈmɒdɪfaɪ] *v.t.* to make modifications

modulate [ˈmɒdjʊleɪt] *v.t.* (*Elec.*) to vary the strength, size or frequency of a sound or radio signal

module [ˈmɒdjul] *n.* (*Eng.*) a standard unit in a number of component parts that fit together to form a machine, building, etc., and which can be detached or moved ▶ The lunar module was released into its own orbit before proceeding to land on the surface of the moon.

molecular [məˈlekjʊlə] *a.* (*Chem.*) having to do with, or made up of, molecules

molecular biology the scientific study of the chemical nature of living organisms,

especially proteins
molecular weight the calculated weight of a molecule of a substance
molecule ['mɒlɪkjul] *n. (Chem.)* the smallest quantity of a substance that can exist without undergoing chemical change
mollusc ['mɒləsk] *n. (Biol.)* an animal belonging to a group of invertebrates, such as snails and oysters, which are protected by a hard shell
momentum [məʊ'mentəm] *n. (Phys.)* a measure of a body's motion ▶ To calculate momentum, you should multiply mass by velocity.
monad ['mɒnæd] *n. (Biol.)* an organism consisting of a single cell
monitor[1] ['mɒnɪtə] *n. (Comput.)* a screen displaying information visually for checking
monitor[2] *v.t.* to observe or control ▶ The progress of the experiment was monitored on closed-circuit television.
mono- *comb. form* single
monochromatic [ˌmɒnə(ʊ)krə'mætɪk] *a. (Phys.: Optics)* having only one colour
monorail ['mɒnə(ʊ)ˌreɪl] *n. (Eng.)* a railway along which trains move on only one rail
moped ['məʊped] *n. (Mech.)* a bicycle which also has a small motor
moraine [mə'reɪn] *n. (Geol.: Geog.)* an area of stones, rocks, and detritus carried down by a glacier and deposited in heaps
morbid ['mɔːbɪd] *a. (Med.)* having to do with disease
morbid anatomy the scientific study of diseased parts of the body
morphine ['mɔːfiːn] *n. (Med.)* a sedative drug derived from opium ▶ Morphine is addictive.
motion ['məʊʃ(ə)n] *n. (Phys.)* movement from place to place
motive force any power that promotes mechanical movement *(also called* **motive power***)*
motor[1] ['məʊtə] *n. (Mech.)* a machine which converts power into movement ▶ The pump was driven by a powerful electric motor.
motor[2] *a. (Biol.)* having to do with producing bodily movements ▶ A motor nerve transmits impulses which stimulate the muscles
mould[1] [məʊld] *n. (Metall.)* a hollow shape into which molten metal or another substance is poured, to form a permanent shape when it solidifies
mould[2] *v.t. (Metall.)* to shape metal etc. by pouring it into a mould
mouse [maʊs] *n. (Comput.)* a device for controlling the movement of a visual indicator on a computer screen
movement ['muːvmənt] *n. (Mech.)* the moving parts of a machine ▶ The clock movement was damaged in the fall and several parts had to be replaced.
mucus ['mjuːkəs] *n. (Biol.)* a sticky liquid, like that produced inside the nose ▶ Snails produce mucus to help them move more easily over the ground.
mulch [mʌltʃ] *n. (Hort.)* a mixture of straw, decaying leaves, grass cuttings, etc., which is placed around the roots of bushes to prevent them from becoming too dry
multi- *(abbr.)* many
multiple ['mʌltɪp(ə)l] *n. (Maths.)* a number which can be divided exactly by a smaller number ▶ 14, 21 and 28 are all multiples of 7.
multiple sclerosis ['mʌltɪp(ə)l skləˈrəʊsɪs] a disease of the nervous system leading to paralysis
multiplication [ˌmʌltɪplɪ'keɪʃ(ə)n] *n. (Maths.)* the adding of a number to itself a specific number of times
multiplier ['mʌltɪplaɪə] *n. (Maths.)* the number of times a number is added to itself ▶ If 20 is multiplied by 12, the multiplier is 12.
multiply[1] ['mʌltɪplaɪ] *v.t. (Maths.)* to add a number to itself a specified number of times
multiply[2] *v.i. (Biol.)* to grow in number by dividing or reproducing ▶ The number of infected cells multiplied rapidly.
muscle ['mʌsl] *n. (Biol.: Med.)* the elastic tissue, attached to the bones of an animal's body, which stretches or contracts as bodily movements are made ▶ After the

unusual exercise, the muscles of their legs were aching and sore.

muscular [ˈmʌskjʊlə] *a. (Biol.: Med.)* concerning muscles, possessing well-developed muscles

muscular dystrophy [ˌmʌskjʊlə ˈdɪstrəfɪ] a disease in which the muscles become progressively weaker

mutant [mjuˈtənt] *a. or n. (Biol.)* an organism that has undergone mutation

mutate [mjuˈteɪt] *v.i (Biol.)* to change by mutation

mutation [mjuˈteɪʃ(ə)n] *n. (Biol.)* any inheritable change in the genes or chromosomes of an organism

myopia [maɪˈəʊpɪə] *n. (Med.)* inability to see things clearly at a distance, short-sightedness

n

nacelle [næˈsel] *n. (Aer.)* the part of an aircraft that houses an engine ▶ Smoke began to pour from the port nacelle, and a warning light came on in the cockpit.

nadir [ˈneɪdɪə] *n. (Astron.)* the point on the Earth's surface opposite to the zenith

naked [ˈneɪkɪd] *a.* uncovered

naked eye the eye working alone, unassisted by the use of a telescope or a microscope ▶ The newly discovered star is too faint to be seen with the naked eye.

naked flame an unprotected flame ▶ The gas should not be exposed to a naked flame or it is liable to explode.

napalm [ˈneɪpɑm] *n. (Chem.)* a highly flammable jelly consisting of petrol and other substances, such as rubber ▶ The blazing napalm from the flamethrower stuck to everything with which it came in contact.

naphtha [ˈnæfθə] *n. (Chem.)* a flammable oil or crystalline substance derived from coal tar ▶ Naphtha is used in manufacturing various products, including explosives.

narco- *comb. form* having to do with numbness or drowsiness

narcosis [nɑˈkəʊsɪs] *n. (Med.)* unconsciousness brought on by narcotic drugs

narcotic [nɑˈkɒtɪk] *n. or a. (Chem.: Med.)* one of a series of drugs, including opium and morphine, which are used to bring about unconsciousness or to relieve pain ▶ The improper use of narcotic drugs may lead to addiction.

narrow [ˈnærəʊ] *a. (Eng.)* built to a limited specification

narrow-gauge railway a railway with less than the standard distance between the lines ▶ A short stretch of narrow-gauge railway was used to haul heavy loads up the steep incline.

narrows [ˈnærəʊz] *n. pl. (Geog.)* the narrow part of a river or channel joining two stretches of water ▶ The boat passed through the narrows with barely a metre to spare on either side.

NASA [ˈnæsə] *(abbrev.)* National Aeronautics and Space Administration – the US space exploration authority

nasal [ˈneɪz(ə)l] *a. (Med.)* having to do with the nose

nasal cavity the hollow space inside the nose

nasal spray a device for inserting a liquid into the nasal cavity ▶ The annual misery of hay-fever can be alleviated by use of a nasal spray.

nascent [ˈnæs(ə)nt] *a. (Chem.)* (of an element or compound) created in atomic form during a chemical reaction ▶ The nascent gases formed during the experiment proved to be very active.

natural [ˈnætʃ(ə)r(ə)l] *a. (Biol.)* having to do with nature

natural gas a gas formed of methane and other elements found trapped below ground ▶ The natural gas found under the North Sea is brought ashore in pipelines.

natural history the study of plants and animals in their natural state

natural science one of the branches of science having to do with the study of the natural world ▶ Biology, Chemistry, Physics and Geology are natural sciences.

natural selection the process by which animals and plants survive by adapting to changes in their environment; the survival of the fittest.

naturalist [ˈnætʃ(ə)rəlɪst] *n. (Biol.)* a person who studies plants and animals in their natural surroundings

nature [ˈneɪtʃə] *n. (Biol.)* the totality of

natural plants and animals, as opposed to man-made things

nausea [ˈnɔzɪə] *n. (Med.)* an uneasy sensation in the stomach which often leads to vomiting ▶ The motion of the ship in the storm brought on an acute attack of nausea.

nauseate [ˈnɔzɪeɪt] *v.t. (Med.)* to sicken or cause to vomit ▶ He was nauseated by the foul smell of rotting vegetation.

nautical [ˈnɔtɪk(ə)l] *a. (Naut.)* having to do with ships or sailors

nautical mile *(Meas.)* a unit of length, used at sea (1,852 metres) ▶ A speed of one nautical mile an hour is a knot.

navigable [ˈnævɪɡəb(ə)l] *a. (Naut.)* (of a channel) through which ships may pass ▶ Most of the northern Russian rivers are navigable only in the summer months.

navigate [ˈnævɪɡeɪt] *v.t. (Naut.)* to direct the passage of a ship or aircraft from one point to another ▶ The vessel was navigated into harbour by the local pilot.

neap [nip] *n. (Naut.)* (of a tide) the lowest point, at the times of the month when the rise and fall are least

nearside [ˈnɪəsaɪd] *n. (Eng.)* the side of a car, etc., nearest to the side of the road on which it is driven

nebula [ˈnebjʊlə] *pl.* **nebulae** [ˈnebjʊli] *n. (Phys.: Astron.)* a hazy cloud of gas or particles visible in the night sky

nebulize [ˈnebjʊlaɪz] *v.t. (Phys.: Mech.)* to convert a liquid into a spray or mist

nebulizer [ˈnebjʊlaɪzə] *n. (Med.)* a device used to nebulize a liquid for inhalation ▶ The use of a nebulizer brings instant relief to sufferers from asthma.

necr(o)- *comb. form* having to do with death

necrobiosis [ˌnekrə(ʊ)baɪˈəʊsɪs] *n. (Biol.)* the natural process of death of the cells

necrophilia [ˌnekrə(ʊ)ˈfɪlɪə] *n. (Psych.)* sexual attraction to dead bodies

necrophobia [ˌnekrə(ʊ)ˈfəʊbɪə] *n. (Psych.)* abnormal fear of death or the dead

needle[1] [ˈnid(ə)l] *n. (Med.)* the part of a hypodermic syringe which may be inserted into a vein or muscle ▶ The use of unsterile needles by drug addicts is a cause of the spread of HIV.

needle[2] *n. (Mech.: Phys.)* the metal arrow of a compass that indicates north and south

needle[3] *n. (Med.)* a a thin metal rod used in acupuncture

needle valve [ˈnid(ə)l ˌvælv] a valve with a needle-shaped moving part

negative[1] [ˈneɡətɪv] *a. (Maths.)* having a value of less than zero

negative[2] *a.* or *n. (Elec)* (of a terminal) having an electrical charge like that of an electron and opposite to that of a proton ▶ Always connect negative to negative, and positive to positive.

negative[3] *n. (Phys.: Photo)* (in photography) a film or plate which has been exposed and developed.

negative feedback the return of part of a mechanical or electronic output to the input ▶ The negative feedback in the audio system made the speaker's announcements totally unintelligible.

neo- *comb. form* new

neo-Darwinism [ˌniə(ʊ)ˈdɑwɪnɪz(ə)m] a modern development of the Darwinian theory of evolution ▶ Neo-Darwinism considers the possibility of inheriting acquired characteristics.

neon [ˈniən] *n. (Chem.)* an inert gaseous element used to make signs and lights

neon light [ˈniənˌlaɪt] a lamp, usually in the form of a long tube, which is filled with neon gas and glows when an electric current is passed through it

neoplasm [ˈniə(ʊ)ˌplæzm] *n. (Biol.)* a malignant growth of tissue in the body

nerve [nɜv] *n. (Med.)* a fibre which conveys messages from the brain to other parts of the body and vice versa

nerve cell [ˈnɜvˌsel] any cell forming part of the nervous system

nerve gas [ˈnɜv ˌɡæs] a gas that acts on the central nervous system, paralysing and often killing anyone who inhales it ▶ The use of nerve gases is prohibited by the Geneva Convention.

nervous [ˈnɜvəs] *a. (Med.)* having to do with

the nerves
nervous system the network of nerve fibres within the body
network [ˈnetwɜk] *n. (Comput.)* a series of interconnected computer terminals ▶ Linking all the computer terminals in the building into a single network improved the efficiency of communication and saved a great deal of time.
neural [ˈnjʊər(ə)l] *a. (Med.)* having to do with the nerves or nervous system
neuralgia [njʊəˈrældʒə] *n. (Med.)* pain in the nerves, especially of the head and face
neuro- *comb. form* having to do with the nerves
neurological [ˌnjʊərəˈlɒdʒɪk(ə)l] *n. (Med.)* having to do with neurology
neurology [njʊəˈrɒlədʒɪ] *n. (Med.)* the study of the characteristics of the nervous system and of nervous disorders
neuron [ˈnjʊərɒn] *n. (Biol.)* a nerve cell
neurosis [njʊəˈrəʊsɪs] *n. (Psych.)* a mild mental disorder ▶ Though most of the time his behaviour was normal, he sometimes displayed symptoms of neurosis and did strange things.
neurotic [njʊəˈrɒtɪk] *n. or a. (Psych.)* suffering from a neurosis
neutral[1] [ˈnjutr(ə)l] *a. (Phys.)* having zero electric charge ▶ The plug will have three wires – one brown (positive), one blue (negative), and one green and yellow (neutral).
neutral[2] *a. (Chem.)* neither acid nor alkaline
neutral[3] *a. or n. (Mech.)* (of the gear of an engine) disconnected from the transmission ▶ Make sure the gear is in neutral before you try to start the engine.
neutralize [ˈnjutrəlaɪz] *v.t. (Chem.: Agric.)* to make neutral ▶ A lime fertilizer will neutralize the acidity of the soil.
neutrino [njuˈtrinəʊ] *n. (Phys.)* an elementary particle that travels at the speed of light
neutron [ˈnjutrɒn] *n. (Phys.)* a small particle of matter that has no electrical charge and combines with a proton to form the nucleus of an atom
neutron bomb a sort of nuclear bomb that destroys life but not property
nickel [ˈnɪkl] *n. (Chem.: Metall.)* a malleable, silvery metallic element that resists corrosion
nickel-plated covered with a thin layer of nickel as protection against rusting ▶ Even after years of use, the nickel-plated surfaces of the machine had not rusted.
nicotine [ˈnɪkətin] *n. (Chem.: Med.)* a poisonous substance found in tobacco ▶ The brown nicotine stains on her fingers gave some idea of the probable state of her lungs.
nipple [ˈnɪpl] *n. (Biol.)* the projecting central part of a mammal's breast, through which milk is sucked
nitrate [ˈnaɪtreɪt] *n. (Chem.)* a salt of nitric acid
nitric [ˈnaɪtrɪk] *a. (Chem.)* containing nitrogen
nitric acid a powerful, yellowish acid used in making fertilizers and explosives
nitrogen [ˈnaɪtrədʒən] *n. (Chem.)* a colourless and odourless gas present in the atmosphere
nitroglycerine [ˈnaɪtrə(ʊ)ˈglɪsərɪn] *n. (Chem.)* a thick, corrosive liquid, made by adding glycerine to nitric and sulphuric acids, which is used in the manufacture of explosives
nodule [ˈnɒdjul] *n. (Med.)* a lump or thickening of tissue in the body ▶ The cause of his hoarseness was found to be a small nodule on one of his vocal cords.
norm [nɔm] *n.* an average level or amount, against which others are measured ▶ For the third month output fell below the norm.
normal [ˈnɔm(ə)l] *a.* conforming with a norm
normalize [ˈnɔməlaɪz] *v.t.* to bring something into conformity with a norm
normative [ˈnɔmətɪv] *a.* (of a programme, etc.) setting out a norm ▶ A normative schedule was laid down for a typical year's production.
nose [nəʊz] *n. (Aer.)* the front part of an aircraft
nose cone [ˈnəʊz ˌkəʊn] the conical forward

section of a missile or space vehicle etc., designed to resist the heat generated on re-entry to the Earth's atmosphere ▶ Special heat-resisting tiles were fixed to the nose cone of the space shuttle.

nose dive [ˈnəʊzˌdaɪv] a sudden dive, with the front section downward ▶ The stricken jet went into a steep nose dive and crashed into the sea.

nose wheel [ˈnəʊzˌwiːl] the wheel fixed or lowered under the nose of an aircraft for landing

nova [ˈnəʊvə] n. (Astron.) a star that suddenly flares and then gradually dies

noxious [ˈnɒkʃəs] a. (Chem.) poisonous or harmful ▶ The air around the chemical plant was polluted with noxious fumes.

nuclear [ˈnjuːklɪə] a. (Phys.) having to do with or involving the nucleus of an atom

nuclear bomb a bomb whose explosive power is caused by nuclear fission or fusion

nuclear energy the energy released during nuclear fission or fusion

nuclear fission the splitting of the nucleus of an atom into roughly equal parts, releasing energy

nuclear fuel the fuel that produces nuclear energy for power stations, etc.

nuclear fusion the fusion of two nuclei, which releases energy

nuclear physics the branch of physics that deals with the structure and behaviour of the nuclei of atoms

nuclear power the electrical etc. power generated by the use of nuclear fuel

nuclear reaction the process of change in structure and energy-content when one nucleus reacts with another

nuclear reactor an installation in which controlled nuclear reaction is produced

nucleic [njuːˈkliːɪk] acid (Chem.: Biol.) one of the two acids (RNA, DNA) that are present in all living cells

nucleus [ˈnjuːklɪəs] pl. **nuclei** [ˈnjuːklɪaɪ] n. (Phys.: Chem.) the central part of an atom ▶ A nucleus has a positive electrical charge, and consists of neutrons, protons and other elementary particles.

numerator [ˈnjuːməreɪtə] n. (Maths.) the part of a fraction above the line ▶ The numerator in $\frac{3}{8}$ is 3.

nutrient [ˈnjuːtrɪənt] n. (Biol.) a mineral substance absorbed by a plant's roots for food or which nourishes an animal

nutriment [ˈnjuːtrɪmənt] n. (Biol.) matter providing nourishment or promoting growth

nutrition [njuːˈtrɪʃ(ə)n] n. (Biol.) the intake and assimilation of matter by plants and animals

nutritious [njuːˈtrɪʃəs] a. (Biol.) nourishing, providing good food

nylon [ˈnaɪlən] n. or a. (Chem.) a synthetic material used in the manufacture of various products ▶ The use of nylon fabrics instead of silk brought about great changes in the textiles markets.

O

obese [əʊˈbiːs] *a. (Med.)* excessively and unhealthily fat

obesity [əʊˈbiːsətɪ] *n. (Med.)* the state of being obese ▶ It is now generally agreed that one of the major causes of heart disease is obesity.

oblate [ˈɒbleɪt] *a. (Maths.)* (of a sphere) flatter at the top and bottom ▶ The Earth is an oblate sphere.

oblique¹ [ə(ʊ)ˈblik] *a. (Maths.)* (of a line) at an angle with another, but not at right angles to it

oblique² *a. (Maths.)* (of an angle) that is greater or smaller than 90°

oblong [ˈɒblɒŋ] *n.* or *a. (Maths.)* a rectangular figure, longer than it is wide

obsolescence [ˌɒbsə(ʊ)ˈles(ə)ns] *n.* the state of being obsolescent

obsolescent [ˌɒbsə(ʊ)ˈles(ə)nt] *a.* becoming out of date ▶ By the time the equipment was delivered, it was already obsolescent.

obsolete [ˈɒbsəliːt] *a.* out of date; no longer in existence ▶ They were still working with obsolete machinery. ▶ The fossilized skeleton showed traces of obsolete wings.

obstetric [ɒbˈstetrɪk] *a. (Med.)* having to do with obstetrics

obstetrician [ˌɒbsteˈtrɪʃ(ə)n] *n. (Med.)* a physician who specializes in obstetrics

obstetrics [ɒbˈstetrɪks] *n. sg. (Med.)* the branch of medicine having to do with childbirth ▶ Recent advances in obstetrics have greatly increased the percentage of live births.

obtuse [əbˈtjuːs] *a. (Maths.)* (of an angle) greater than 90°

octagon [ˈɒktəgən] *n. (Maths.)* a flat, eight-sided figure

octahedron [ˌɒktəˈhiːdr(ə)n] *n. (Maths.)* a solid figure with eight faces

octane [ˈɒkteɪn] *n. (Chem.)* the hydrocarbon compound in petrol, in terms of which its quality is graded ▶ The low-octane petrol used in many countries has a very distinctive smell.

ocular [ˈɒkjʊlə] *a. (Med.)* relating to the eye

ocularist [ˈɒkjʊlərɪst] *n. (Phys.: Med.)* a person who makes artificial eyes

oculist [ˈɒkjʊlɪst] *n. (Med.)* an old term for ophthalmologist *or* optician.

odometer [əʊˈdɒmɪtə] *n. (Mech.)* an instrument that measures the distance a vehicle has travelled ▶ In a car, the odometer is more often called a mileometer.

oedema [iˈdiːmə] *n. (Med.)* a swelling caused by the accumulation of fluid

oesophagus [iˈsɒfəgəs] *n. (Med.)* the tube in the throat down which food passes to the stomach

offside [ˈɒfsaɪd] *a.* or *n. (Mech.)* (of a vehicle) the side normally furthest from the side of the road ▶ The car was stopped by the police because its offside rear light was not working.

ohm [əʊm] *n. (Elec.: Meas.)* a unit of electrical resistance ▶ 1 ohm is a resistance of 1 amp at a pressure of 1 volt.

ohmmeter [ˈəʊmˌmiːtə] *n. (Elec.)* an instrument for measuring electrical resistance in ohms

oil [ɔɪl] *n. (Chem.)* an alternative term for petroleum

oil-bearing [ˈɔɪlˌbeərɪŋ] (of rock) containing oil

oil-fired [ˈɔɪlˌfaɪəd] (of a boiler, etc.) fuelled by some form of oil

oil gauge [ˈɔɪlˌgeɪdʒ] an instrument for measuring the level of oil in a tank ▶ The oil gauge indicated that the level of oil was so low that the engine was in danger of overheating.

oil rig [ˈɔɪl ˌrɪg] a structure for drilling for oil,

especially in the seabed ▶ Constructing an oil rig in the North Sea was an extremely difficult feat of engineering.

oil slick [ˈɔɪl ˌslɪk] a mass of oil floating on the sea ▶ The tanker ran aground and a huge oil slick formed, killing hundreds of seabirds..

oil well [ˈɔɪl ˌwel] a hole bored in the ground, from which oil is extracted

oilfield [ˈɔɪlfild] *n.* an area where oil has been discovered and is being extracted

oleaginous [ˌɒlɪˈædʒɪnəs] *a. (Biol. : Geol.)* oily, fatty

olfactory [ɒlˈfæktərɪ] *a. (Biol.)* having to do with the sense of smell

olfactory nerve the nerve linking the brain and the nose ▶ Sniffing industrial fumes can be harmful to the olfactory nerve.

omni- *comb. form* all

omnidirectional [ˌɒmnɪdaɪˈrekʃ(ə)n(ə)l] *a. (Radio)* (of an aerial) receiving radio signals from all directions

omnivore [ˈɒmnɪvɔ] *n. (Biol.)* a creature that eats food of all kinds, whether of vegetable or of animal origin

omnivorous [ɒmˈnɪvərəs] *a. (Biol.)* having the eating habits of an omnivore ▶ By nature, Man is omnivorous, but many people choose to become vegetarians.

on line *(Comput.)* (of data) stored in and accessible from a computer ▶ All the laboratory's research data is now on line.

opacity [əʊˈpæsətɪ] *n. (Phys.)* the quality of being opaque

opaque [əʊˈpeɪk] *a. (Phys.)* not translucent, not allowing light to pass through ▶ Opaque areas on the X-ray photographs showed the presence of infected tissue.

open [ˈəʊpən] *a.* not closed or enclosed

open-cast [ˈəʊpənˌkɑst] carried out from the surface, without sinking a shaft ▶ A huge chunk had been carved out of the mountainside by open-cast mining for diamonds.

open circuit an incomplete or broken electrical circuit

open-heart (of surgery) relating to an operation on the heart, usually with artificial control of the blood circulation

open hearth (of a furnace) having a shallow hearth in which scrap or pig-iron is heated, usually by gas

operable [ˈɒpərəb(ə)l] *a. (Med.)* capable of being the subject of a surgical operation

operate¹ [ˈɒpəreɪt] *v.t. (Mech.)* to drive a machine or bring it into operation

operate² *v.i. (Med.)* to perform a surgical operation ▶ The surgeon requested the patient's written permission to operate.

operating table [ˈɒpəreɪtɪŋ ˌteɪb(ə)l] a high, flat bed on which a patient lies for a surgical operation

operating theatre [ˈɒpəreɪtɪŋ ˌθɪətə] a specially equipped room in which surgical operations are performed

operation [ɒpəˈreɪʃ(ə)n] *n. (Med.)* repair or removal by surgery of part of the body

operational [ɒpəˈreɪʃ(ə)n(ə)l] *a.* having to do with, ready for, or capable of action

operational research analysis of a problem in order to design a model to solve it

operator [ˈɒpəreɪtə] *n. (Mech.)* someone who operates a machine

ophthalmologist [ˌɒfθælˈmɒlədʒɪst] *n. (Med.)* a specialist in diagnosing or treating diseases of the eye

ophthalmology [ˌɒfθælˈmɒlədʒɪ] *n. (Med.)* the branch of medicine having to do with the eye or with diseases of the eye

ophthalmoscope [ɒfˈθælməskəʊp] *n. (Med.)* an instrument for examining the eye

opiate [ˈəʊpɪət] *n. (Med.)* a drug containing opium

opium [ˈəʊpɪəm] *n. (Med.)* a narcotic drug made from poppy seeds

optic [ˈɒptɪk] *a. (Med.)* relating to the eye or to vision ▶ Pressure on the optic nerve may cause blurred vision.

optical [ˈɒptɪk(ə)l] *a. (Phys.)* relating to the eye or to sight ▶ What the audience thought they saw was really an optical illusion.

optician [ɒpˈtɪʃ(ə)n] *n. (Med.)* someone who makes spectacles or other optical instruments ▶ You should visit an optician regularly to ensure that your spectacles are

suited to your eyes.
optics ['ɒptɪks] *n. sg. (Phys.)* the branch of science which deals with vision and light
orbit[1] ['ɔbɪt] *n. (Phys.)* the path followed by an electron around the nucleus of an atom
orbit[2] *n. or v.t. (Maths.: Astron.)* the path followed by a satellite around the Earth or another heavenly body, or to follow such a path ▶ The satellite went into orbit as planned and will continue to orbit the Earth for several decades.
orbital ['ɔbɪt(ə)l] *a.* having to do with an orbit or a route around something ▶ The new orbital road around the city is already overcrowded with traffic.
order ['ɔdə] *n. (Biol.)* one of the taxonomic groups under which plants and animals are classified
ordinal ['ɔdɪn(ə)l] *a. (Maths.)* having a position in a sequence of numbers
ordinal number a number indicating a position in a group or sequence ▶ First, second, third, etc. are ordinal numbers, but 1, 2, 3, etc. are cardinal numbers.
ore [ɔ] *n. (Geol.: Metall.)* rock from which metals can be extracted
organ ['ɔgən] *n. (Biol.: Med.)* a part of a plant or animal having a specific function ▶ The vital organs of a human body include the heart, lungs, liver, kidneys, etc.
organic [ɔ'gænɪk] *a. (Chem.)* (of a compound) which contains carbon; having to do with living things or their organs ▶ Organic chemistry deals with living things, and inorganic chemistry with inanimate things. ▶ The space probe revealed no signs of organic matter on the moon.
organism ['ɔgənɪz(ə)m] *n. (Biol.)* a living being ▶ All around us there are millions of tiny organisms which are invisible to the naked eye.
orgasm ['ɔgæz(ə)m] *n. (Biol.)* the culmination of sexual excitement
orifice ['ɒrɪfɪs] *n. (Biol.)* an opening or hole in the body ▶ The two nasal orifices are called nostrils.
ornithology [ˌɔnɪθɒlədʒɪ] *n. (Biol.)* the scientific study of birds

ortho- *comb. form* straightness, uprightness
orthodontics [ˌɔθə(ʊ)'dɒntɪks] *n. sg. (Med.)* the branch of dentistry having to do with making teeth straight or regular
orthopaedics [ˌɔθə'pidɪks] *n. sg. (Med.)* the branch of surgery having to do with skeletal problems
oscillate ['ɒsɪleɪt] *v.i. (Phys.: Mech.)* to swing like a pendulum, regularly alternating from one position to another
oscillator ['ɒsɪleɪtə] *n. (Elec.)* an instrument that creates an alternation of electric current
oscillogram [ə'sɪlə(ʊ)græm] *n. (Elec.: Mech.)* a recording or trace of an oscillation visible on the screen of an oscillograph
oscillograph [ə'sɪləgraf] *n. (Elec.: Mech.)* a machine for recording electrical or other oscillations on paper
oscilloscope [ə'sɪlə(ʊ)skəʊp] *n. (Elec.: Mech.)* a machine for displaying oscillations on a screen
ossiferous [ɒ'sɪfərəs] *a. (Geol.)* (of rock) containing bones, or of the sort in which bones are often found
ossify ['ɒsɪfaɪ] *v.i. (Biol.: Geol.)* to turn into bone or fossil ▶ Clutches of ossified dinosaur eggs have been found in the Gobi desert.
out [aʊt] *a. or advb.* not in, beyond the limits of a specific time, place, etc.
outboard ['aʊtbɒd] *a. (Mech.)* (of a motor) fitted on the stern of a boat, outside the boat itself ▶ An outboard motor is easy to instal or remove from a boat.
outcrop ['aʊtkrɒp] *n. (Geol.)* a part of a layer of rock visible above the surface of the ground ▶ The prospectors found several outcrops of gold-bearing rock in the hills.
outer space the universe beyond the Earth's atmosphere ▶ The damaged satellite left its orbit and disappeared into outer space.
outfall ['aʊtfɔl] *n. (Eng.: Ecol.)* the end of a sewer or drain that empties its contents into something ▶ The outfall from the factory drains discharged toxic waste into the river and poisoned all the fish.
outlet ['aʊtlɪt] *n. (Elec.)* a point in an electric

circuit at which current can be tapped ▶ In the laboratory are numerous outlets into which pieces of apparatus may be plugged.

output¹ [ˈaʊtpʊt] *n. (Elec.)* the electric current delivered from a circuit or device ▶ The output of the battery was measured on a voltmeter.

output² *n. (Comput.)* the amount of data that a computer can produce ▶ The output of the latest generation of computers is much greater than that of earlier models.

oval [ˈəʊv(ə)l] *n.* or *a. (Maths.)* egg-shaped, or elliptical

ovarian [əʊˈveərɪən] *n. (Biol.)* having to do with the ovaries

ovary [ˈəʊvərɪ] *n. (Biol.)* one of the female organs in which eggs are produced

overdose [ˈəʊvəˌdəʊs] *n.* or *v.i. (Med.)* too great a dose of a drug, etc., or to take such a dose ▶ He became very depressed, and took an overdose of sleeping pills in order to draw attention to himself.

overload [ˈəʊvələʊd or ˌəʊvəˈləʊd] *n.* or *v.t. (Elec.)* an excessive demand, e.g. for current, to make an excessive demand ▶ So many appliances were switched on at the same time that the system became overloaded and broke down.

oxide [ˈɒksaɪd] *n. (Chem.)* a compound of oxygen with another element

oxidize [ˈɒksɪdaɪz] *v.i.* and *v.t. (Chem.)* to form or cause to form a compound with oxygen ▶ The copper leads were so heavily oxidized that the current had stopped flowing.

oxyacetylene [ˌɒksɪəˈsetəlin] *n. (Chem.: Mech.)* a mixture of oxygen with acetylene which burns with a very hot flame ▶ The thieves burned away the lock of the safe with a powerful oxyacetylene lamp.

oxygen [ˈɒksɪdʒən] *n. (Chem.)* a colourless, odourless gaseous element which forms almost half the Earth's atmosphere and is essential for life ▶ Oxygen combines with hydrogen to form water (H_2O).

oxygen mask [ˈɒksɪdʒən ˌmɑːsk] a device worn over the nose and mouth, through which oxygen is passed ▶ Pilots flying at high altitudes had to wear oxygen masks to assist their breathing.

oxygen tent [ˈɒksɪdʒən ˌtent] a sealed enclosure around a hospital bed into which oxygen is pumped to help the patient to breathe

oxygenate [ɒkˈsɪdʒəneɪt] *v.t. (Chem.)* to oxidize or impregnate with oxygen

ozone [ˈəʊzəʊn] *n. (Chem.)* a colourless gas formed when an electrical charge is passed through oxygen

ozone layer a concentration of ozone high in the Earth's atmosphere that filters out the Sun's ultra-violet rays ▶ Ecologists are becoming increasingly worried at the appearance of holes in the ozone layer.

p

pacemaker [ˈpeɪsˌmeɪkə] *n.* (*Med.*) an electronic device surgically implanted in the chest to regulate the heartbeat

package [ˈpækɪdʒ] *n.* (*Comput.*) a set of ready-made computer programs designed to deal with a specific problem or topic ▶ Without the appropriate package of programs, a computer is of only limited use.

pad [pæd] *n.* (*Aer.*) an area for take-off and landing, especially for helicopters

paediatrician [ˌpidɪəˈtrɪʃ(ə)n] *n.* (*Med.*) a doctor who specializes in paediatrics

paediatrics [ˌpidɪˈætrks] *n. sg.* (*Med.*) the branch of medicine that deals with children's diseases

pan [pæn] *v.i.* (*Photo.*) (of a video or film camera) to move laterally in order to get a panoramic view ▶ The camera panned slowly over the plain, following the herd of antelope.

pan- *comb. form* all-inclusiveness

panchromatic [ˌpænkrəˈmætɪk] *a.* (*Photo.*) (of film) sensitive to all colours ▶ Since the invention of panchromatic film, colour photographs have familiarized millions with sights they will never themselves see.

pancreas [ˈpænkrɪəs] *n.* (*Med.*) a gland near the stomach that secretes a fluid which aids digestion

pandemic [pænˈdemɪk] *a.* (*Med.*) (of a disease) affecting people over a wide geographical area ▶ Some diseases which were pandemic in Europe a century ago are now rarely seen.

para- *comb. form* nearness

paraffin [ˈpærəfɪn] *n.* (*Chem.*) a derivative of petroleum, used as a fuel for heating

parallel [ˈpærəlel] *a.* or *n.* (*Maths.*) (of lines) being the same distance apart at all points

parallel of latitude any imaginary line on a map of the world's surface connecting all points at the same distance north or south of the equator ▶ The border between Canada and the USA runs along the 49th parallel of latitude.

parallelogram [ˌpærəˈlelə(ʊ)græm] *n.* (*Maths.*) a four-sided figure with two pairs of parallel sides

paralyse [ˈpærəlaɪz] *v.t.* (*Med.*) to make insensitive or unable to move

paralysis [pəˈræləsɪs] *n.* (*Med.*) the state of being paralysed ▶ After his stroke he suffered partial paralysis of his left arm and leg.

paramedic(al) [ˌpærəˈmedɪk(ə)l] *n.* (*Med.*) a person trained to supplement the work of a doctor ▶ A team of paramedics was assisting in the treatment of the victims of the crash.

parameter [pəˈræmɪtə] *n.* (*Maths.*) an agreed value given to a symbol used in a mathematical formula, e.g. $x = 1$ ▶ Without knowing the parameters, we cannot solve the problem.

paranoia [ˌpærəˈnɔɪə] *n.* (*Psych.*) a mental disorder often involving unreasonable fears

paranoid [ˈpærənɔɪd] *a.* suffering from paranoia ▶ After he had been questioned by the police, he became totally paranoid and was sure he was being followed wherever he went.

paraplegia [ˌpærəˈplidʒə] *n.* (*Med.*) paralysis of the lower half of the body

paraplegic [ˌpærəˈplidʒɪk] *a.* or *n.* (*Med.*) suffering from paraplegia

parasite [ˈpærəsaɪt] *n.* (*Biol.: Med.*) an animal or plant that lives on another one and feeds on it ▶ Ivy is a common parasite that eventually kills the tree on which it lives.

parasitology [ˌpærəsaɪˈtɒlədʒɪ] *n. (Biol.)* the branch of science that deals with the study of parasites

part [pɑt] *n. (Mech.)* a modular piece of a machine or apparatus which is manufactured separately for assembly with other parts ▶ Some of the old ironworks have been replaced by light industries manufacturing electronic parts.

particle [ˈpɑtɪk(ə)l] *n. (Phys.)* a piece of matter so small that it is considered to have no significant size or structure

particle accelerator a machine used in nuclear physics to accelerate electrically charged particles to very high energies

parturition [ˌpɑtjʊ(ə)ˈrɪʃ(ə)n] *n. (Biol.)* the process of giving birth

passive [ˈpæsɪv] *a. (Chem.)* (of a substance) which will not react chemically

passive smoking exposure to poisoning by inhaling another person's tobacco smoke

pasteurization [ˌpɑstjəraɪˈzeɪʃ(ə)n] *n. (Biol.)* the process of heating a liquid in order to kill any germs or to slow down the process of fermentation

pasteurize [ˈpɑstjəraɪz] *v.t. (Biol.)* to subject milk or another liquid to the process of pasteurization ▶ Pasteurized milk may be stored for a long period without going sour.

patch [pætʃ] *v.t. (Radio)* to connect together several points by the use of temporary circuits ▶ Despite the chaos after the storm, the telephone operator managed to patch the journalist through to his office.

patent [ˈpeɪtənt] *n.* a licence or permission to sell, guaranteeing the quality of the item sold

patent medicine a medicinal preparation offered for sale without a doctor's prescription

pathological [ˌpæθəˈlɒdʒɪk(ə)l] *a. (Med.)* having to do with pathology or with a disease

pathology [pæˈθɒlədʒɪ] *n. (Med.)* the branch of medicine that deals with the cause, nature and origin of diseases ▶ Specialists were unable to decide on a treatment until they had studied the pathology of the mysterious disease.

payload [ˈpeɪləʊd] the part of a cargo for which a charge can be made, or the weight of bombs, etc., which can be carried ▶ The weight of extra fuel needed for the journey decreased the value of the payload considerably. ▶ Using less heavy fuel increases the payload.

pedometer [peˈdɒmɪtə] *n. (Mech.)* an instrument which registers the number of steps taken in order to calculate the distance walked

pelvis [ˈpelvɪs] *n. (Biol.)* the arrangement of bones to which the hind legs of animals are attached

pendulum [ˈpendjʊləm] *n. (Mech.)* the oscillating device that activates the mechanism of a clock

penicillin [ˌpenɪˈsɪlɪn] *n. (Chem.: Med.)* a powerful antibiotic used to combat infection ▶ The chance discovery of penicillin has made formerly fatal infections possible to cure.

pentagon [ˈpentəgən] *n. (Maths.)* a flat figure with five sides ▶ Probably the best example is the American military building (the Pentagon) in Washington.

perigee [ˈperɪdʒi] *n. (Astron.)* the point in its orbit at which the Moon or a satellite is nearest to the Earth

period[1] [ˈpɪərɪəd] *n. (Astron.)* the time taken by a body such as the Earth to complete one turn about its axis

period[2] *n. (Geol.)* a specific span of time in the history of the Earth

period[3] *n. (Maths.: Phys.)* the time taken to complete one cycle in a process that consists of a number of regular cycles, such as the swing of a pendulum or oscillation of current

period[4] *n. (Biol.)* a monthly flow of blood from the uterus of a woman of childbearing age

period[5] *n. (Chem.)* one of the rows of elements in the periodic table

periodic table a table in which chemical elements are arranged according to their atomic weights

peripheral [pəˈrɪf(ə)r(ə)l] *n. (Comput.)* a part of a computer system connected to the main device but not an integral part of it ▶ Peripherals include the disk, mouse, etc. ▶ The main elements in a computer system are the hardware, the software, and the peripherals.

permafrost [ˈpɜməfrɒst] *n. (Geol.)* a layer of ground which is permanently frozen

permutation [ˌpɜmjʊˈteɪʃ(ə)n] *n. (Maths.)* one of several possible arrangements of a set of numbers ▶ The possible permutations of the numbers on the combination lock made it almost impossible for an unauthorized person to guess the combination.

perpendicular [ˌpɜpənˈdɪkjʊlə] *a.* or *n. (Maths.)* a line drawn at an angle of 90° to the horizontal

pest [pest] *n. (Biol.)* a plant or animal that harms or destroys crops ▶ Creatures regarded as pests may in fact have an important role in the ecological chain.

pesticide [ˈpestɪsaɪd] *n. (Chem.: Agric.)* a chemical preparation for killing pests, especially insects

petrify [ˈpetrɪfaɪ] *v.i. (Geol.)* to turn into stone through a process of fossilization ▶ A forest of petrified trees was unearthed in the desert.

petro- *comb. form* having to do with petroleum or rocks

petrochemical [ˌpetrəʊˈkemɪk(ə)l] *n. (Chem.)* (of a chemical) derived from petroleum

petrol [ˈpetr(ə)l] *n. (Chem.)* a flammable liquid obtained by refining petroleum

petroleum [pəˈtrəʊlɪəm] *n. (Chem.)* a mineral oil trapped underground and extracted through oil wells for the production of paraffin, petrol, lubricating oil, etc.

pharmaceutical [ˌfɑməˈsjutɪk(ə)l] *a. (Chem.: Med.)* having to do with pharmacy or medicinal drugs

pharmacist [ˈfɑməsɪst] *n. (Chem.: Med.)* a specialist who prepares drugs and medicines

pharmacologist [ˌfɑməˈkɒlədʒɪst] *n. (Chem.:*

Med.) a specialist in pharmacology

pharmacology [ˌfɑməˈkɒlədʒɪ] *n. (Chem.: Med.)* the study of the use of drugs in medicine

pharmacy [ˈfɑməsɪ] *n. (Chem.: Med.)* the study of the preparation of drugs for use in medicine, or the place where the drugs are distributed

pharynx [ˈfærɪŋks] *n. (Biol.)* the cavity behind the mouth and above the larynx

phase^1 [feɪz] *n. (Astron.)* (of the Moon and planets) the appearance determined by the amount of lighted surface visible at a given time ▶ In its first phase, the Moon appears as a crescent.

phase^2 *n. (Phys.)* the stage reached in any periodic process at a given time ▶ The first phase of the building was completed ahead of schedule.

phase^3 *n. (Elec.)* one of the electrical circuits in a multi-circuit system

phial [ˈfaɪəl] *n. (Med.)* a sealed glass tube containing one dose of a medicine or drug

phonetics [fəˈnetɪks] *n. sg. (Phys.)* the study of the sounds of human speech

phosphate [ˈfɒsfeɪt] *n. (Chem.)* any salt or compound of phosphoric acid, especially one used as a fertilizer ▶ The addition of phosphates and nitrates has greatly increased the productivity of the land.

phosphorescence [ˌfɒsfəˈres(ə)ns] *n. (Chem.)* the quality of giving out light without heat

phosphoric [fɒsˈfɒrɪk] *a. (Chem.)* having to do with high-valency phosphorus

phosphorus [ˈfɒsf(ə)rəs] *n. (Chem.)* a pale, highly flammable, non-metallic element that glows in the dark ▶ Phosphorus is important in the manufacture of matches.

photo- *comb. form* having to do with light

photocopy [ˈfəʊtəˌkɒpɪ] *v.t.* or *n. (Phot.: Mech.)* to make a copy of a document by a photographic process, such as xerox; the copy so made

photoelectric cell a device using the effect of light to generate an electric current ▶ The popular name for a photoelectric cell is 'electric eye'.

photolithography [ˌfəʊtə(ʊ)ˈlɪθɒɡrafi] *n.*

(*Photo.: Mech.*) the transference of an image to a metal plate by photographic means as a basis for printing

photosensitive [ˌfəʊtə(ʊ)ˈsensətɪv] *a.* (*Phys.*) reacting to light ▶ Negatives are printed on photosensitive paper.

photostat [ˈfəʊtə(ʊ)stæt] *n.* (*Photo.: Mech.*) a photocopy

photosynthesis [ˌfəʊtə(ʊ)ˈsɪnθəsɪs] *n.* (*Bot.*) the process by which green plants use sunlight to convert carbon dioxide and water into food

physician [fɪˈzɪʃ(ə)n] *n.* (*Med.*) a medical doctor who does not normally perform surgery

physicist [ˈfɪzɪsɪst] *n.* a specialist in physics

physics [ˈfɪzɪks] *n. sg.* the branch of science that deals with the physical properties of matter and energy ▶ Students of physics learn about such things as heat, light and sound.

physio- *comb. form* nature and natural functions

physiological [ˌfɪzɪə(ʊ)ˈlɒdʒɪk(ə)l] *a.* (*Med.*) having to do with the body and its functions

physiology [ˌfɪzɪˈɒlədʒɪ] *n.* (*Med.*) the branch of science concerned with the functions of living bodies ▶ Physiology is an important branch of medicine.

physiotherapy [ˌfɪzɪə(ʊ)ˈθerəpɪ] *n.* (*Med.*) exercise and treatment, especially after illness or surgery, to restore such bodily functions as the use of muscles and joints ▶ The operation on the patient's leg was a complete success, but he needed a long course of physiotherapy before he could walk again.

pier [pɪə] *n.* (*Build.*) a pillar supporting a bridge or archway ▶ The central pier was swept away in the floods, and the bridge collapsed into the river.

pig [pɪg] *n.* (*Metall.*) a moulded mass of smelted iron which has not been purified or worked ▶ Sailing ships sometimes used pig-iron as ballast.

pigment [ˈpɪgmənt] *n.* (*Biol.*) a substance in plants, animals and humans that gives them a specific colouring

pigmentation [ˌpɪgmenˈteɪʃ(ə)n] *n.* (*Biol.*) the coloration of skin and other matter

pile¹ [paɪl] *n.* (*Elec.*) a dry battery for generating electric current ▶ The technical name for an atomic reactor is 'atomic pile'.

pile² *n.* (*Build.*) a column of concrete or other material driven into the ground ▶ Excavations in the city revealed traces of wooden piles from buildings which disappeared long ago.

pile driver a mechanical hammer for driving piles into the ground

pinion [ˈpɪnjən] *n.* (*Mech.*) the smaller of two cog-wheels that mesh together

pink [pɪŋk] *v.i.* (*Mech.*) (of a petrol engine) to make a knocking sound because of faulty fuel supply

pipe [paɪp] *n.* (*Mech.*) a hollow tube for carrying fluids or gases

pipeline [ˈpaɪpˌlaɪn] *n.* (*Mech.*) a connected series of pipes for conveying liquid or gas over a long distance ▶ Oil from the Arabian Gulf is carried by pipeline to Turkey.

pistil [ˈpɪstɪl] *n.* (*Bot.*) the female organ of flowering plants

piston [ˈpɪstən] *n.* (*Mech.*) part of an engine, consisting of a metal disc which fits closely inside a cylinder and is driven up and down by the explosion of the ignited mixture

piston ring [ˈpɪst(ə)n ˌrɪŋ] a ring fitted tightly around a piston to ensure that it fits snugly into the cylinder

piston rod [ˈpɪst(ə)n ˌrɒd] a rod attached to a piston that moves up and down with it and is connected to other moving parts ▶ Another name for piston rod is 'connecting rod'.

pitch¹ [pɪtʃ] *n.* (*Build.*) a sticky, black substance derived from coal-tar or occurring naturally ▶ Pitch is soft when hot and hard when cold, and is used for surfacing roads.

pitch² *n.* (*Build.*) the angle of the slope of a roof ▶ The pitch of chalet roofs is made very steep so that the snow slides off before it becomes too thick and heavy.

pitch³ *n. or v.t. (Phys.)* the high or low degree of a musical note or a voice ▶ His high-pitched voice could clearly be heard above the general hubbub in the room.

pith [pıθ] *n. (Bot.)* a spongy substance found in the stems of some plants and in the peel of some fruits ▶ They removed the pith from lengths of reeds to make whistles and pipes.

pivot [ˈpıvət] *n. (Mech.)* a central point about which something swings or turns ▶ The pivot of the swing bridge in the harbour was a central pier, on which it turned to open or close.

pivotal [ˈpıvət(ə)l] *a. (Mech.)* having to do with the point or pivot about which something turns

placebo [pləˈsibəʊ] *n. (Med.)* a substance, which has no observable physical effect, prescribed as a medicine in order to give a patient psychological comfort ▶ Placebos are also used with control groups in experiments to test new drugs.

placenta [pləˈsentə] *n. (Med.)* the organ through which the foetus is fed inside a mammal's womb

plane¹ [pleın] *n. (Maths.)* a flat surface ▶ A straight line drawn along a plane will touch it at all points.

plane² *n. (Mech.: Aer.)* the flat surface which, when inclined and moved forward, provides the lift which enables an aircraft to fly ▶ Aircraft have main planes (wings) and tailplanes.

planet [ˈplænıt] *n. (Astron.)* a heavenly body revolving around the Sun ▶ The major planets are Mercury, Venus, Mars, Jupiter and Saturn.

planetarium [ˌplænıˈteərıəm] *n. (Mech.: Astron.)* a building in which the movements of the planets are demonstrated on a dome representing the heavens

plankton [ˈplæŋ(k)tən] *n. (Biol.)* tiny forms of animal and plant life found in water, especially in the sea ▶ Destruction of plankton by the dumping of toxic waste can have serious effects on other forms of marine life.

plant [plɑnt] *n. (Eng.)* a large factory or industrial installation

plasma¹ [ˈplæzmə] *n. (Phys.: Astron.)* a gas in which particles with negative and positive charges are present in roughly equal numbers ▶ The Sun and most stars contain plasma.

plasma² *n. (Med.)* the yellowish liquid in which blood corpuscles float ▶ Processed blood plasma is used in blood transfusions.

plaster [ˈplɑstə] *n. (Build.)* a mixture of lime, sand and water, applied as a paste to walls and ceilings and forming a smooth surface when dry

plaster cast [ˌplɑstə ˈkɑst]: a cast of plaster of Paris, used to hold a broken limb immobile as it mends

plaster of Paris a white powder which can be mixed with water to form a paste which can then be moulded ▶ Plaster of Paris hardens and retains its shape when it dries.

plasterboard [ˈplɑstəˌbɔd] a thin board consisting partly of plaster, used in making walls and ceilings

plastic¹ [ˈplæstık] *n. (Chem.)* a durable synthetic substance which can be moulded under heat or pressure ▶ Plastic is now used to make objects which used to be made of wood, lead, etc.

plastic² *a.* malleable; capable of being moulded

plastic surgery a form of surgery which repairs damaged parts of the body by replacing them with tissue taken from elsewhere in the body ▶ Severe burns can now be treated very effectively through plastic surgery.

plasticity [plæsˈtısıtı] *n. (Chem.)* (of a substance) the capability of being moulded when soft and retaining its shape when hardened

plate [pleıt] *v.t. (Metall.)* to coat one metal with a layer of another, usually silver, gold or nickel ▶ The exposed surfaces of the machine were plated with chrome to prevent them from rusting.

play [pleɪ] n. (Mech.) room for movement in a combination of tightly fitted moving parts of a machine ▶ Too much play between the piston and cylinder wall leads to loss of power.

plug¹ [plʌg] n. (Mech.) a piece of wood, rubber or other material which can be used to block up a hole or gap and which can be removed and replaced ▶ When the plug was removed, the liquid quickly ran out of the tank.

plug² n. (Elec.) a normally plastic device used to attach electric leads to an outlet by inserting positive, negative and earth pins into a socket in a wall, etc.

plug³ (in) v.t. (Mech.: Elec.) to insert a plug in a socket

plumb [plʌm] a. (Mech.) precisely vertical

plumbline [ˈplʌmlaɪn] n. (Mech.) a device for establishing whether a wall is plumb by dangling a weight on the end of a line so that it hangs vertically

plunge [plʌndʒ] v.i. and v.t. (Mech.) to thrust downward, usually into a liquid, such as water

plunger [ˈplʌndʒə] n. (Mech.) a moving machine part which operates with a plunging motion ▶ Depressing the plunger will detonate the explosive charge.

pneumatic [njuˈmætɪk] a. (Phys.: Mech.) (of a mechanism) operated by air pressure

pneumatic drill a hand-held, pressure-powered machine for drilling holes in road surfaces

pneumatic tyre a tyre filled with air under pressure

pneumonia [njuˈməʊnɪə] n. (Med.) acute inflammation of the lungs

point¹ [pɔɪnt] n. (Geog.: Maths.) one of the 32 marks on the circumference of a compass indicating specific directions ▶ The cardinal points are north, south, east and west.

point² n. (Elec.: Mech.) one of the contacts in a petrol engine between which the electric spark leaps in order to ignite the mixture ▶ If the points are corroded, the current will not flow and the engine will not start.

point³ n. (Mech.) a switch by which railway or tram lines may be connected or disconnected to change the direction in which a vehicle moves ▶ In heavy frost the points may freeze and trains or trams cannot then be directed along the correct lines.

point⁴ v.t. (Build.) to finish a new wall by pressing and trimming the mortar or cement between courses of bricks etc. ▶ Walls should be properly pointed to keep out the damp.

polar¹ [ˈpəʊlə] a. (Geog.) having to do with or near the North or South Poles

polar² (Elec.) having to do with a positive or a negative electric pole

polarity [pə(ʊ)ˈlærɪtɪ] n. (Elec.) being either a negative or a positive pole

pole¹ [pəʊl] n. (Geog.) one of the extremities of the imaginary axis about which the Earth is supposed to turn ▶ Both North and South Poles are permanently covered with ice.

pole² n. (Elec.) one of the two terminals, e.g. of an electric battery, which have either positive or negative polarity

poliomyelitis [ˌpəʊlɪə(ʊ)maɪəˈlaɪtɪs] n. (Med.) an infectious viral disease that affects the central nervous system ▶ He walked with a limp due to paralysis resulting from poliomyelitis.

pollen [ˈpɒlən] n. (Bot.) the powder discharged by male organs of flowers, which fertilizers female organs ▶ Pollen is often carried from flower to flower by bees.

pollen count [ˈpɒlən ˌkaʊnt] a measure of the amount of pollen present in the air

pollutant [pəˈluːt(ə)nt] n. (Ecol.) waste matter, often industrial, which pollutes the atmosphere, water or soil ▶ Pollutants from the factory stacks blighted the trees for miles around.

pollute [pəˈluːt] v.t. (Ecol.) to make impure through the release of toxic waste etc.

pollution [pəˈluːʃ(ə)n] n. (Ecol.) the process or result of polluting

poly (abbr.) polytechnic

poly-¹ prefix containing a polymer

poly-² comb. form many

polyclinic [ˈpɒlɪˌklɪnɪk] n. (Med.) a clinic able to treat many different kinds of diseases

polyester [ˌpɒlɪˈestə] *n.* *(Chem.)* one of the polymer synthetic materials used for making plastics

polygon [ˈpɒlɪgən] *n.* *(Maths.)* a flat figure with more than five sides

polyhedron [ˌpɒlɪˈhidrən] *n.* *(Maths.)* a solid figure with more than five planes

polymer [ˈpɒlɪmə] *n.* *(Chem.)* a chemical compound with a simple structure of large molecules

polytechnic [ˌpɒlɪˈteknɪk] *a.* or *n.* of a college where advanced courses are taught in a wide range of subjects

polythene [ˈpɒlɪθin] *n.* *(Chem.)* a very pliable form of plastic commonly used to make bags

polyunsaturated [ˌpɒlɪʌnˈsætʃəreɪtɪd] *a.* *(Med.)* (of fats) having a specific molecular structure which does not foster the formation of cholesterol in the blood ▶ Polyunsaturated fats are considered more suitable in food than other fats.

polyurethane [ˌpɒlˈjʊərəθeɪn] *n.* *(Chem.)* a form of synthetic plastic commonly used in packaging

polyvinyl chloride see **PVC**

pontoon [pɒnˈtun] *n.* *(Build.)* a flat-bottomed boat used in bridge-building

pontoon bridge a temporary bridge laid on floating pontoons ▶ The main bridge had been destroyed in the bombing, so the supplies were taken across the river on a pontoon bridge.

pore[1] [pɔ] *n.* *(Biol.)* a small hole in the skin, hide, or surface of a person, animal, or plant, through which liquid or gas may escape ▶ Perspiration dripped from every pore as the climbers struggled to the summit of the mountain.

pore[2] *n.* *(Geol.)* a small hole or space in a rock surface

porous [ˈpɔrəs] *a.* *(Biol.: Geol.)* having, or having to do with, pores

port[1] [pɔt] *n.* *(Naut.)* a watertight doorway in the side of a ship, through which cargo, etc., may be loaded or unloaded

port[2] *n.* *(Mech.)* an aperture in the cylinder of a petrol engine, through which lubricating oil is admitted

port[3] *n.* *(Comput.)* a circuit for the input or output of data

port[4] *n.* *(Naut.: Aer.)* the left side ▶ Navigators refer to port and starboard instead of left and right.

porthole [ˈpɔthʊl] *n.* *(Naut.)* a watertight window in the side of a ship

positive[1] [ˈpɒzətɪv] *a.* *(Maths.)* having a value greater than zero ▶ The Chairman forecast a positive balance.

positive[2] *a.* *(Elec.)* having an electrical charge like that of a proton, and opposite to that of an electron

positive[3] *n.* *(Photo.)* a print made from a photographic plate or negative

post- *comb. form* after, following

postcode [ˈpəʊs(t)kəʊd] *n.* a combination of letters and numbers denoting a subsection of a postal area, and used as an aid in sorting mail ▶ The postcode of the London office is W1A 2AB.

postmortem [ˌpəʊs(t)ˈmɔtəm] *a.* *(Med.)* after death

postmortem examination the medical examination of a body after death ▶ If there is any doubt about the cause of death, a postmortem examination is held.

postnatal [ˌpəʊs(t)ˈneɪt(ə)l] *a.* *(Med.)* having to do with the period immediately after the birth of a baby ▶ The young mother and child were given regular postnatal care.

postoperative [ˌpəʊstˈɒpərətɪv] *a.* *(Med.)* having to do with the period just after a surgical operation

potash [ˈpɒtæʃ] *n.* *(Chem.)* a powerful alkali obtained from the ashes of plants ▶ Potash in the form of wood ash is used by gardeners to neutralize acid soil.

potential [pəʊˈtenʃl] *n.* *(Elec.)* the energy of an electric charge, measured in volts

potential difference the difference in state between two points in an electric field which makes the current flow from one to the other

potentiometer [pəˌtenʃɪˈɒmɪtə] *n.* *(Elec.)* an instrument for measuring differences in electric potential

power¹ [ˈpaʊə] *n.* *(Maths.)* the value of a number multiplied by itself ▶ $2 \times 2 \times 2 \times 2 \times 2 = 2^5$, or 2 to the power of 5 ($=32$).

power² *n.* *(Phys.)* mechanical energy, as opposed to manual work

power³ *n.* *(Elec.)* the strength of an electric current, measured in watts

power cut [ˈpaʊə ˌkʌt] a temporary break in power supply, usually as a result of mechanical failure

power drill [ˈpaʊə ˌdrɪl] an electrically-driven tool for boring holes in wood or metal

power house [ˈpaʊə ˌhaʊs] an electricity-generating plant

power line [ˈpaʊə ˌlaɪn] a cable conveying high-voltage electric power

power pack [ˈpaʊə ˌpæk] a device linked into an electrical circuit to convert direct current to alternating current

power point [ˈpaʊə ˌpɔɪnt] a wall socket into which an electric plug may be inserted

power station [ˈpaʊə ˌsteɪʃ(ə)n] an electricity-generating plant

power supply [ˈpaʊə səˌplaɪ] an electrical supply line into which appliances, etc., may be plugged

pre- *comb. form* before, in advance of

precast [ˌpriˈkɑst] *a.* *(Build.)* cast in a desired form before use

precast concrete ready-cast pillars or beams of concrete for use in building construction

precipitate [prɪˈsɪpɪteɪt] *v.i.* and *v.t.* *(Chem.)* to separate or cause to separate as a solid substance from a solution ▶ The impurities had to be precipitated and removed from the solution before the experiment could begin.

precipitation [prɪˌsɪpɪˈteɪʃ(ə)n] *n.* *(Meteor.)* falls of hail, rain, sleet, or snow, caused by the condensation of water vapour in the atmosphere

preclinical [ˌpriˈklɪnɪk(ə)l] *a.* *(Med.)* (of a disease) in a very early stage of development, before accurate diagnosis can be attempted ▶ At the preclinical stage, the symptoms can be quite misleading.

predator [ˈpredətə] *n.* *(Biol.)* an animal that feeds on others ▶ The most vicious and dangerous predator is Man.

prefabricated [ˌpriˈfæbrɪkeɪtɪd] *(abbr.* **prefab**) *a.* (Build.) (of materials) manufactured beforehand and assembled on the spot ▶ Prefabricated houses erected fifty years ago are still in use.

pregnancy [ˈpregnənsɪ] *n.* *(Med.)* (of mammals) the state of having a foetus growing in the uterus

pregnant [ˈpregnənt] *a.* *(Med.)* being in the condition of pregnancy

premature [ˈpremətʃə] *a.* *(Med.)* (of a baby) born before the normal full term of pregnancy has expired ▶ The premature child was so weak that it had to be kept at first in an incubator.

premedical [ˌpriˈmedɪk(ə)l] *n.* or *a.* *(Med.)* a preliminary course of study undertaken before beginning the study of medicine proper

premedication [ˌpriˌmedɪˈkeɪʃ(ə)n] *n.* *(Med.)* medical treatment in preparation for a surgical operation

prescribe [prɪˈskraɪb] *v.t.* *(Med.)* (of a doctor) to give directions for treatment or medicine

pressure [ˈpreʃə] *n.* *(Phys.)* the force exerted on one body by another with which it is in contact ▶ The swollen waters exerted immense pressure on the wall of the dam.

pressure cabin [ˈpreʃə ˌkæbɪn] the cabin of an aircraft, spacecraft, etc., in which the air pressure is maintained at a high level

pressure cooker [ˈpreʃə ˌkʊkə] an airtight pot oven for rapid cooking at high temperature and pressure

pressure point [ˈpreʃə ˌpɔɪnt] one of the points on the body where a blood vessel may be pressed against a bone in order to stop bleeding

prestress [ˌpriˈstres] *v.t.* *(Build.)* to subject material beforehand to the sort of pressure it will undergo in use

prestressed concrete concrete pillars, etc., reinforced with metal rods already stretched by being subjected to the sort of load they will encounter in actual use ▶ The use of prestressed concrete makes it possible to build much higher.

primary ['praɪmərɪ] *a.* being the earliest or most basic of its kind

primary cell a cell that converts chemical energy into electrical energy

primary colour one of the basic colours, from which all other colours may be mixed ▶ The primary colours are red, yellow and blue.

prime¹ [praɪm] *a.* original, primary

prime² *v.t. (Mech.)* to fill a machine, such as a pump, with fluid in order to expel all the air ▶ It was necessary to prime the pump before it could be started up.

prime³ *v.t. (Mech.)* (of a petrol engine) to inject fuel into the carburettor so that the mixture may be ignited and the engine started ▶ There is a small plunger that may be used to prime the carburettor.

prime⁴ *v.t. (Mech.)* (of a cannon, etc.) to put in place a small charge to be used in setting off the main charge ▶ The guns were already primed when the order came to withdraw.

prime⁵ *(Chem.: Build.)* to prepare new wood for painting ▶ The new doors and windows were primed with a substance which prevented the paint from being absorbed.

prime number [ˌpraɪm ˈnʌmbə] a number that can only be divided by 1 or by itself ▶ 1, 3, 5 etc. are prime numbers.

primitive ['prɪmətɪv] *a. (Biol.: Mech.: etc.)* relating to an earlier stage of development

primitive animal an invertebrate creature such as a jellyfish, an octopus or an earthworm ▶ Shellfish are classified as primitive animals; spiders and insects are not.

print [prɪnt] *v.t. (Mech.)* to make an impression on paper or another substance by pressing against it a surface covered with ink or some other matter

printed circuit an electrical circuit printed in metal on a thin board of insulating material ▶ A faulty printed circuit can easily be taken out and replaced.

printer ['prɪntə] *n. (Comput.)* one of the peripherals of a computer system, by which information loaded in the memory may be printed on paper ▶ The three main kinds of printer are the matrix, the daisy-wheel, and the laser printer.

printout ['prɪntaʊt] *n. (Comput.)* data printed on paper from a computer through the printer

prism ['prɪzm] *n. (Phys.)* a transparent object, usually with a number of parallel sides, which is used to break down light into its constituent colours

probe [prəʊb] *n. (Med.)* a thin metal instrument for examining inside a wound

process ['prəʊses] *v.t. (Comput.)* to perform the necessary operations on the data loaded into a computer in order to obtain the required information ▶ The research findings were processed through the mainframe computer and printed out for distribution.

processor ['prəʊsesə] *n. (Comput.)* the central processing unit of a computer

product ['prɒdʌkt] *n. (Maths.)* the result of multiplying two or more numbers ▶ The product of 10 × 10 is 100.

progeny ['prɒdʒənɪ] *n. (Biol.)* offspring, children

prognosis [prɒgˈnəʊsɪs] *pl.* **prognoses** [prɒgˈnəʊsiːz] *n. (Med.)* a prediction of the likely development of a disease or condition ▶ At such an early stage the physician hesitated to make a firm prognosis.

program ['prəʊgræm] *n. (Comput.)* a set of coded instructions fed into a computer instructing it to perform specified operations on the data loaded into it ▶ The adoption of the new program made it possible to process the data in many more ways.

programmer ['prəʊgræmə] *n. (Comput.)* someone who analyses problems and designs appropriate computer programs

progression [prə(ʊ)ˈgreʃ(ə)n] *n. (Maths.)* a sequence of numbers, in which each number has a fixed relationship to the previous one ▶ Progressions may be arithmetical (e.g. 2, 4, 6, 8) or geometrical (e.g. 2, 4, 8, 16).

proof¹ [pruf] *n. (Maths.)* a series of operations demonstrating that a statement is true ▶ You should learn the proof of Pythagoras's theorem.

proof² *n. (Photo.)* a trial print made from a negative ▶ You will get a number of proofs of the photograph so that you can select which to use for the positives.

proof³ *n. (Chem.)* the strength of an alcoholic drink measured in degrees on a scale of 1–100 ▶ Spirits have a higher proof than wines.

proof⁴ *a. (Chem.)* having been made resistant to something ▶ The new plastic material was proof against damage by heat. ▶ It was also waterproof.

propagate [ˈprɒpəgeɪt] *v.t. (Biol.)* to increase the number of plants or animals from parent stock ▶ Shrubs not native to the country are sometimes difficult to propagate.

propel [prəˈpel] *v.t. (Mech.)* to cause something to move forward ▶ The little boat was propelled across the bay by a strong crosswind.

propellant [prəˈpelənt] *n. (Chem.: Aer.)* a fuel or gas used in rockets and aerosols

propeller [prəˈpelə] *n. (Mech.: Naut.: Aer.)* a rotating device, with two or four blades, which is driven by an engine and provides the motive force for a boat or aircraft

proper [ˈprɒpə] *a.* belonging to one person or entity

proper fraction a fraction with a value less than 1 ▶ In a proper fraction the numerator (above the line) is smaller than the denominator (below the line).

property [ˈprɒpətɪ] *n. (Chem.)* a constant and unchanging characteristic of a substance ▶ An important topic in chemistry is the properties of matter.

prophylactic [ˌprɒfɪˈlæktɪk] *n.* or *a. (Med.)* a substance intended to prevent disease ▶ Efficient prophylactic medicine reduces the resources required for treatment of the sick.

propulsion [prəˈpʌlʃ(ə)n] *n. (Mech.)* (of an engine) the driving force ▶ The invention of jet propulsion has revolutionized air travel.

prospect [prɒsˈpekt] *v.i. (Geol.)* to search for ores or deposits of minerals, especially gold and precious stones

prospector [prəˈspektə] *n. (Geol.)* someone who prospects for ores ▶ Many prospectors searched in vain for many years, while others struck rich deposits by accident.

prosthesis [ˌprɒsˈθiːsɪs] *n. (Med.)* an artificial part of the body, usually a limb

protein [ˈprəʊtiːn] *n. (Biol.)* a body-building substance present in many types of food ▶ Eggs, meat and fish are rich in protein.

proto- *comb. form* original, first

proton [ˈprəʊtɒn] *n. (Phys.)* an elementary particle, with a positive electrical charge, which is present in the nucleus of an atom

protoplasm [ˈprəʊtə(ʊ)ˌplæz(ə)m] *n. (Biol.)* a colourless jelly, present in all living cells and tissue

psych-, psyche-, psycho- *comb. form* having to do with the mind

psychedelic [ˌsaɪkəˈdelɪk] *a. (Psych.)* having to do with the effect of taking hallucinogenic drugs ▶ Taking drugs like LSD produces psychedelic illusions which can be very dangerous.

psychiatry [saɪˈkaɪətrɪ] *n. (Med.)* the branch of medicine that deals with mental disorders and their treatment ▶ Many people confuse psychiatry, which is a branch of medicine, with psychology, which is not.

psychoanalysis [ˌsaɪkə(ʊ)əˈnæləsɪs] *n. (Psych.)* a method of treating the mentally or emotionally disordered by a system of questioning that reveals what is in the patient's subconscious mind

psychology [saɪˈkɒlədʒɪ] *n.* the study of the mind and how it works

psychopath [ˈsaɪkə(ʊ)ˌpæθ] *n. (Psych.)* someone suffering from a mental disorder which often leads to acts of violence and a failure to feel guilt at having committed them

psychotherapy [ˌsaɪkə(ʊ)ˈθerəpɪ] *n. (Med.)* the treatment of mental disorders by psychological techniques

pulley ['pʊlɪ] *n. (Mech.)* a device for lifting heavy loads ▶ A pulley consists of a fixed, grooved wheel, over which a rope is passed, to one end of which the load is attached, while the other end is pulled in order to raise or lower the load to another level.

pulsar ['pʌlsɑ] *n. (Astron.)* a small, dense star, which rotates very quickly and sends out radio waves and other forms of radiation in regular pulses

pulse^1 [pʌls] *n. (Med.)* the changing of pressure of the blood against the wall of an artery as the heart beats ▶ Though the patient had lost a lot of blood, a faint pulse indicated that he was still alive.

pulse^2 *n. (Elec.)* a sudden change in an electric current

pumice ['pʌmɪs] *a.* or *n. (Geol.)* a light and porous rock formed by volcanic lava ▶ Pumice stone may be used as a mild abrasive, e.g. for scrubbing hands.

pump^1 [pʌmp] *n. (Mech.)* a device for forcing in or sucking out water or air ▶ When the boat sprang a leak, all the crew were called upon to man the pumps.

pump^2 *v.t. (Mech.)* to make use of a pump ▶ More air had to be pumped into the tyres.

punch [pʌntʃ] *n. (Mech.)* a device for making holes in metal, paper, etc.

punched cards *(Comput.)* one of the methods employed in primitive data retrieval systems, in which cards bearing items of data have holes punched in them in specific positions along the top, so that all those with holes in a certain place can be extracted mechanically without sorting through the remainder

puncture^1 ['pʌŋktʃə] *v.t. (Mech.)* to punch a hole, e.g. in an inflated tyre ▶ The sharp stones in the rough track punctured our tyres several times.

puncture^2 *n. (Mech.)* a hole in an inflated tyre, etc.

pupil ['pjup(ə)l] *n. (Biol.)* the circular area at the centre of the eye through which light passes

purchase ['pɜtʃəs] *n. (Mech.)* a firm grip or hold on a nearly slippery surface ▶ Because of the oil spilled on the road, the tyres were unable to gain a firm purchase.

purity ['pjʊərətɪ] *n. (Chem.)* freedom from pollution by other substances

pus [pʌs] *n. (Med.)* a yellowish liquid secreted from inflamed or infected wounds

PVC [ˌpiviˈsi] *(abbr.)* polyvinyl chloride *(Chem.)* a synthetic material used in the manufacture of numerous items, including clothing, pipes and electrical insulating materials

pylon ['paɪlən] *n. (Eng.)* a tall, steel tower supporting high-tension electric cables

pyramid ['pɪrəmɪd] *n. (Math.)* a solid figure, usually with a square base, which has flat, triangular sides that meet in a point at the top ▶ The pyramids of ancient Egypt are still among the wonders of the world.

pyro- *comb. form* having to do with fire

pyromagnetic [ˌpaɪrə(ʊ)mægˈnetɪk] *a. (Phys.)* changing in magnetic intensity according to temperature

q

quadrangle [ˈkwɒdræŋg(ə)l] *n. (Maths.)* a plane figure with four sides and four angles

quadrant [ˈkwɒdrənt] *n. (Maths.)* a quarter of a circle, or of its circumference ▶ A quadrant is an arc of 90°.

quadraphonics [ˌkwɒdrəˈfɒnɪks] *n. sg. (Phys.)* a system of making and distributing sound recording by the use of four inputs (e.g. microphones) and four loudspeakers

quadratic [kwɒˈdrætɪk] *a. (Maths.)* involving a squared number, e.g. 5^2 (=25)

quadratic equation an equation using squared powers, e.g. x^2, y^2, etc.

quadri- *comb. form* four

quadrilateral [ˌkwɒdrɪˈlæt(ə)r(ə)l] *a.* or *n. (Maths.)* four-sided; a flat figure with four sides ▶ Quadrilaterals include squares, rectangles, parallelograms, etc.

quadruped [ˈkwɒdrʊped] *n. (Biol.)* an animal with four legs ▶ Most mammals are quadrupeds.

quadruple [ˈkwɒdrup(ə)l] *v.i.* and *v.t. (Maths.)* to multiply by four ▶ The crops were quadrupled by the application of fertilizers to the land.

quantity [ˈkwɒntətɪ] *n. (Maths.)* a definable amount

quantum [ˈkwɒntəm] *n. (Phys.)* the smallest quantity of energy that a system can possess

quantum statistics the statistics of the distribution of elementary particles or atoms

quantum theory the theory that the energy of electrons in radiation is in units that cannot be divided into smaller ones

quarantine [ˈkwɒrənˌtin] *n.* a period of compulsory isolation for people or animals feared to be carrying an infectious disease ▶ All pet animals brought into the UK must be put into quarantine for six months.

quark [kwɑk] *n. (Phys.)* one of the very small parts of which elementary particles are made up

quasar [ˈkweɪzɑ] *n. (Phys.: Astron.)* a very distant star or other body that emits intense radiation

quaternary¹ [kwəˈtɜnərɪ] *a. (Chem.)* consisting of an atom bound to four other atoms or groups

quaternary² *a. (Maths.)* having four variables

quench¹ [kwentʃ] *v.t. (Phys.)* to reduce the activity of molecules in giving off light

quench² *v.t. (Elec.)* to reduce the number of sparks given off when a circuit is cut off

quicklime [ˈkwɪklaɪm] *n. (Chem.)* burned lime which has not been slaked

quintuple [kwɪnˈtjʊp(ə)l] *v.t. (Maths.)* to multiply by five

quotient [ˈkwəʊʃ(ə)nt] *n. (Maths.)* the result of dividing one number by another ▶ If 10 is divided by 2, the quotient is 5

r

rabid ['ræbɪd] *a. (Zool.)* (of an animal) suffering from rabies

rabies ['reɪbiz] *n. (Zool.: Med.)* a nervous disease which causes animals to foam at the mouth and may cause madness and death to humans if transmitted by the bite of a rabid animal

race [reɪs] *n. (Eng.)* a strong and fast flow of water caused by restricting the width of the stream or channel through which it runs ▶ The rush of water in the mill-race turned the huge wheel round.

rack [ræk] *n. (Mech.)* a bar with teeth around or along it that engage in those of a cogwheel to make it turn ▶ The rack and wheel together are called 'rack and pinion'.

rack railway a small railway, usually up a steep slope, along which carriages are drawn by means of a rack and pinion mechanism

radar ['reɪdɑ] *n. (Radio)* a means of detecting the presence of solid objects, such as aircraft or ships, by bouncing radio waves off them ▶ The use of radar makes it much safer to navigate in conditions of poor visibility.

radar screen a small screen, like a TV screen, on which the objects detected by radio are made visible ▶ A line of dots on the left of the radar screen indicated the approach of a number of planes.

radial ['reɪdɪəl] *a. (Maths.)* having to do with a circular arrangement with spokes from the centre to the circumference, like a bicycle wheel

radial engine a petrol engine with a number of cylinders arranged radially ▶ Early propeller-driven aircraft had radial engines.

radial tyre a pneumatic tyre in which the grooves on the surface are arranged in a radial pattern in order to give better purchase

radian ['reɪdɪən] *n. (Maths.)* the angle between two radii crossing the circumference of a circle the same distance apart as their length (57·296°)

radiance ['reɪdɪəns] *n. (Phys.)* the quality of being radiant

radiant ['reɪdɪənt] *a. (Phys.)* sending out light or heat

radiate ['reɪdɪeɪt] *v.t.* and *v.i. (Phys.)* to send out light or heat

radiation [ˌreɪdɪ'eɪʃ(ə)n] *n. (Phys.)* the process or result of emitting light or heat

radiation sickness the illness caused by exposure to radioactivity ▶ The leakage from the atomic reactor led to an outbreak of radiation sickness among the workers.

radiator[1] ['reɪdɪeɪtə] *n. (Mech.)* a device for radiating heat, either by passing hot water through pipes, or by passing an electric current through material with high resistance

radiator[2] *(Mech.)* a means of cooling an engine by passing a fluid in pipes through it, cooling them in the current of air generated by forward motion or by a fan, and then recirculating the fluid through the pipes

radical[1] ['rædɪk(ə)l] *n. (Phys.: Chem.)* a group of atoms which exists unchanged in a number of different compounds and can therefore be treated as a single atom

radical[2] *(Maths.)* the square root of a quantity ▶ If the radical of 9 is 3, and of 16 is 4, what is the radical of 25?

radio ['reɪdɪəʊ] *n. (Phys.)* a means of sending or receiving signals through the air by electro-magnetic waves ▶ What used to be called 'wireless' is now usually called radio.

radio astronomy the use of a radio telescope to receive and analyse radio signals sent out by bodies in space

radio beacon a radio installation that sends out a fixed signal as an aid to navigation by ships and aircraft ▶ Radio beacons have the same sort of function as lighthouses.

radio beam the fixed signal sent out by a radio beacon

radio compass a device that gives directional bearings by receiving and interpreting radio signals, such as radio beams

radio control control of a mechanism from a distance by the use of radio waves ▶ The robot which examines suspected bombs is operated by radio control.

radio frequency the wave frequency on which a radio signal is transmitted

radio signal a signal emitted by radio

radio telescope a telescope used to pick up and analyse radio signals from outer space ▶ By means of a radio telescope, astronomers can receive signals from heavenly bodies far beyond the visual range of optical telescopes.

radioactive [ˌreɪdɪəʊˈæktɪv] *a. (Phys.)* having to do with radioactivity ▶ Certain kinds of rock, occurring naturally, are now known to be radioactive.

radioactive fallout radioactive dust etc., which results from an atomic explosion ▶ The radioactive fallout from the Chernobyl disaster was detected as far away as Wales.

radioactivity [ˌreɪdɪəʊækˈtɪvətɪ] *n. (Phys.)* invisible rays, harmful to life, caused by atomic fission

radiography [ˌreɪdɪˈɒɡrəfɪ] *n. (Med.)* the production of X-rays for use in medical diagnosis

radiotherapy [ˌreɪdɪə(ʊ)ˈθerəpɪ] *n. (Med.)* the use of X-rays or radioactive substances in the treatment of diseases

radium [ˈreɪdɪəm] *n. (Phys.)* a radioactive metallic element used in radiotherapy

radius [ˈreɪdɪəs] *pl.* **radii** [ˈreɪdɪaɪ] *n. (Maths.)* a straight line from the centre of a circle to a point on the circumference

rainforest [ˈreɪnˌfɒrɪst] *n. (Biol.: Ecol.)* a tropical forest where rainfall is heavy and the trees grow close together ▶ Ecologists are worried about the increasing destruction of rainforests, especially in Brazil.

raise [reɪz] *v.t. (Maths.)* to multiply a number by itself a given number of times ▶ 3 raised to the power of 5 (3^5) = $3 \times 3 \times 3 \times 3 \times 3$ (= 243).

rake [reɪk] *n. (Naut.)* the slope of a ship's superstructure ▶ The lookout identified the approaching vessel by the rake of her funnels.

RAM [ræm] *(abbr.) n. (Comput.)* random access memory – the temporary storage space on a computer which is lost when the machine is switched off

ramp [ræmp] *n. (Naut.)* a slope, sometimes mobile, enabling wheeled vehicles to move from one level to another ▶ The ferry lowered the ramp and the cars drove slowly down on to the quay.

range[1] [reɪndʒ] *n.* an area of land set aside for the firing or testing of weapons, such as artillery or small arms ▶ A red flag is flown when the range is in use.

range[2] *(Mech.: Phot.)* the distance between someone firing a gun or taking a photograph and the target, or the distance at which a gun or a camera lens can be effective ▶ By the time the cannon was primed and ready, the target was already out of range.

range-finder [ˈreɪndʒˌfaɪndə] a device for measuring the distance between a gun or a camera and the target ▶ The adoption of a new design of range-finder greatly improved the performance.

re- *comb. form* repetition, renewal

react [riˈækt] *v.i. (Chem.)* to undergo a change when in contact with another substance ▶ The acid reacted with the marble chips to give off a foul-smelling gas.

reaction [riˈækʃ(ə)n] *n. (Chem.: Phys.)* the process of chemical change ▶ The addition of a reagent set off a chain reaction.

reactor [riˈæktə] *n. (Phys.)* a power station

using nuclear energy to generate electricity

read [riːd] *v.t. (Comput.)* (of a system) to accept data coded in a specific form ▶ The new system can read any commonly used computer language.

reagent [rɪˈeɪdʒənt] *n. (Chem.)* an element which, when added to other substances, sets off a chemical reaction

receiver [rɪˈsiːvə] *n. (Radio)* a device which can accept incoming electromagnetic signals and convert them into sound or visual images ▶ A powerful new receiver was installed and made it possible to receive signals from almost anywhere in the world.

reception [rɪˈsepʃ(ə)n] *n. (Radio)* the process of receiving signals, or the quality of the sound or images resulting ▶ During the tropical storm reception was so bad that we could hardly make out what the speakers were saying.

reciprocate [rɪˈsɪprəkeɪt] *v.i. (Mech.)* (of an engine part) to move backward and forward in a straight line ▶ The piston of a petrol engine reciprocates in the cylinder.

reciprocating engine [rɪˈsɪprəˌkeɪtɪŋ ˌendʒɪn] an engine driven by a mechanism that reciprocates

reclaim [rɪˈkleɪm] *v.t. (Agric.: Ecol.)* (of land) to bring back into productive use after flooding, pollution or neglect ▶ Land laid waste by outworked industrial processes is now being reclaimed for agricultural purposes.

reclamation [ˌrekləˈmeɪʃ(ə)n] *n. (Agric.: Ecol.)* the process of restoring land to use ▶ A major programme of reclamation is now proposed for the old coal-mining areas.

recoil [rɪˈkɔɪl] *n. (Mech.)* a sudden backward movement of a gun after firing ▶ The violent recoil of the old rifles made accurate aiming very difficult.

reconstitute [rɪˈkɒnstɪtjuːt] *v.t.* to restore something to its original form ▶ Many dried foods may be reconstituted by the addition of water.

rectangle [ˈrektæŋgl] *n. (Maths.)* a flat, four-sided figure with rightangles

rectify [ˈrektɪfaɪ] *v.t. (Elec.)* to change the flow of an electric current from alternating to direct

rectilinear [ˌrektɪˈlɪnɪə] *a. (Maths.)* in a straight line

recto [ˈrektəʊ] *n. or a.* the right hand page of an open book ▶ The title of the book was printed on the recto, but the verso was blank.

recur [rɪˈkɜː] *v.i. (Maths.)* (of a decimal number) to be repeated forever ▶ The decimal expression of $\frac{1}{3} = 0\cdot33$ recurring.

recycle [ˈriːsaɪk(ə)l] *v.t. (Ecol.)* to process something already used, so that it may be used again ▶ Paper, bottles, aluminium cans, etc., are now regularly recycled, thus saving energy and other resources.

reduce[1] [rɪˈdjuːs] *v.t. (Maths.)* to simplify a fraction, equation, or other expression by using simpler terms ▶ The fraction $\frac{6}{3}$ can be reduced to 2, and $1\frac{1}{2}$ can be reduced to 1·5.

reduce[2] *v.t. (Med.)* to manipulate a displaced or fractured part of the body back into position ▶ The climber's fractured leg was reduced and bound in splints.

reduce[3] *v.t. (Chem.)* to change a compound by removing oxygen from it in a chemical reaction ▶ The water in the tank was reduced to hydrogen by the process of hydrolysis.

reducing agent [rɪˈdjuːsɪŋ ˌeɪdʒənt] *(Chem.)* a substance used to produce a chemical reaction, which removes oxygen from compounds by combining with them to form oxides

re-entry [rɪˈentrɪ] *n. (Aer.: Mech.)* the act of re-entering the Earth's atmosphere from space ▶ The tiles on the nose cone of the space shuttle are specially designed to resist the intense heat generated on re-entry.

refine [rɪˈfaɪn] *v.t. (Chem.)* to purify a substance such as sugar, or to separate a substance such as petroleum into its constituent parts

refinery [rɪˈfaɪnərɪ] *n. (Chem.)* an installation at which the process of refining is carried out ▶ The crude oil is conveyed from the oil wells to the refinery by means of a pipeline.

refit [ˌriːˈfɪt] v.t. or n.(Naut.) (of a ship) to overhaul and repair for future use ▶ After her battering in the gales she had to go into dry dock to be refitted. ▶ After her refit, she put to sea again.

reflect [rɪˈflekt] v.t. (Phys.) to throw back heat, light or sound ▶ The intense heat of the sun was reflected off the metal surface. ▶ The outline of the trees was perfectly reflected in the still water of the lake. ▶ The sound of her voice was reflected by the cliffs.

reflex [ˈriːfleks] a. (Maths.) (of an angle) greater than 180°

refract [rɪˈfrækt] v.t. (Phys.) to cause a beam of light to seem to change direction on passing through the surface of water, etc., at an angle ▶ The beam of the searchlight was refracted by a thin layer of cloud.

refrigerant [rɪˈfrɪdʒərənt] n. (Phys.: Chem.) a substance such as carbon dioxide used as a coolant in a refrigerator ▶ Releasing refrigerants into the atmosphere damages the ozone layer.

refrigerate [rɪˈfrɪdʒəreɪt] v.t. (Phys.: Mech.) to store substances at low temperatures in order to preserve them ▶ Food may now be kept for long periods by being refrigerated.

refrigerator [rɪˈfrɪdʒəreɪtə] n. (Phys.: Mech.) a box-shaped container in which goods are preserved at low temperatures

regenerate [riˈdʒenəreɪt] v.t. and v.i. (Zool.) to renew itself ▶ The severed tail of an earthworm will regenerate itself.

register [ˈredʒɪstə] v.i. (Mech.) to show up on a calibrated scale ▶ The earthquake registered 6·5 on the Richter scale.

regulate [ˈregjʊleɪt] v.t. (Mech.) to make a mechanism work correctly or accurately ▶ The clock was regulated to keep perfect time.

regulator [ˈregjʊleɪtə] n. (Mech.) a device that automatically regulates a mechanism

reinforce [riːɪnˈfɔːs] v.t. to strengthen one sort of material by adding another

reinforced concrete concrete strengthened by having iron bars embedded in it ▶ A skeleton of reinforced concrete supported the weight of the building.

rejig [ˌriːˈdʒɪg] v.t. (Eng.) to re-equip with modern machinery ▶ The old factory was completely rejigged with state-of-the-art technological equipment.

relativity [ˌreləˈtɪvətɪ] n. (Phys.) the relationship between time, size and mass ▶ Einstein's Theory of Relativity revolutionized scientific thought.

relay [ˈriːleɪ] n. (Radio) a mechanism that receives radio or telephone signals and passes them on ▶ A series of relays conveyed the signals over vast distances.

relict [ˈrelɪkt] a. (Biol.) persisting after the species in general has become extinct ▶ Scientists working in the Gobi desert came across relict forms of plant life not found anywhere else in the world.

remainder [rɪˈmeɪndə] n. (Maths.) what is left after the division of one number into another into which it will not go exactly ▶ $7 \div 3 = 2$, remainder 1.

remote [rɪˈməʊt] a. distant, far-off

remote control operation of a vehicle or other mechanism from a distance, usually by means of radio signals ▶ The bomb was detonated by remote control.

renal [ˈriːn(ə)l] a. (Med.) having to do with the kidneys ▶ After sampling the local brandy she suffered an acute attack of renal colic.

render[1] [ˈrendə] v.t. (Chem.) (of fat) to purify by melting and removing unwanted matter ▶ The fat from the carcasses was rendered for use in making soap.

render[2] v.t. (Build.) to cover a wall etc. with a thin skin of plaster ▶ The walls were rendered and decorated with fresh paint.

repeater [rɪˈpiːtə] n. (Radio) a device that increases the strength of incoming radio signals ▶ The installation of a repeater improved radio reception throughout the area.

repeating [rɪˈpiːtɪŋ] a. (Mech.) (of a mechanical

action) one which repeats itself without having to be reset

repeating rifle [rɪˈpitɪŋ ˌraɪf(ə)l] a rifle that can be fired a number of times without being reloaded

replicate [ˈreplɪkeɪt] v.i. and v.t. (Biol.) (of a cell) to reproduce, make copies of itself ▶ The cells replicate themselves more quickly in certain circumstances.

reproduce [ˌriprəˈdjus] v.i. (Biol.) to produce young of the species ▶ Mammals reproduce by giving birth to live young.

reproduction [ˌriprəˈdʌkʃ(ə)n] n. (Biol.) the process of reproducing ▶ The rate of reproduction of the rabbit is very high.

reproductive [ˌriprəˈdʌktɪv] a. (Biol.) having to do with reproduction

reproductive organs the organs of the body used for the process of reproduction

reproductive system the method by which reproduction is achieved ▶ The reproductive system of many plants makes use of the wind to scatter seeds.

reprography [rɪˈprɒgrəfɪ] n. (Mech.) a system of reproducing copies of printed sheets, e.g. by xerox

reptile [ˈreptaɪl] n. (Zool.) one of the class of crawling animals ▶ Reptiles include snakes, lizards, crocodiles and turtles.

reserve [rɪˈzɜv] n. (Ecol.) an area set aside in order to preserve natural animal or plant life ▶ Areas of China where the panda breeds have been declared special nature reserves.

reservoir [ˈrezəvwɑ] n. (Eng.: Mech.) an area or receptacle in which water, oil or some other fluid is collected

resin [ˈrezɪn] n. (Bot.: Chem.) a sticky substance exuded by some plants and trees, or synthetized by a chemical process ▶ Resin dripped down the trunk of the pine tree.

resistance [rɪˈzɪst(ə)ns] n. (Elec.) the ability of a substance to resist the passage through it of an electric current ▶ Metals such as copper have a very low resistance.

resistor [rɪˈzɪstə] n. (Elec.) a piece of wire inserted into an electrical circuit in order to increase its resistance

respiration [ˌrespəˈreɪʃ(ə)n] n. (Biol.) the act of breathing

respirator [ˈrespəreɪtə] n. (Mech.) a mask for the nose and mouth to facilitate breathing in atmospheres where there are noxious fumes, etc.

respiratory [rɪˈspɪrət(ə)rɪ] a. (Biol.: Med.) having to do with respiration

respiratory disease a disease affecting the organs used in breathing ▶ The most widespread respiratory disease is the common cold.

respond [rɪˈspɒnd] v.i. (Med.) to show a change in status following a course of treatment ▶ Though powerful drugs were used, the patient failed to respond and grew rapidly worse.

retort [rɪˈtɔt] n. (Chem.) a conical bottle used in laboratory experiments

retread [ˌriˈtred] v.t. (Mech.) (of a tyre) to cover with a new layer of rubber, preparatory for further use

retrieval [rɪˈtriv(ə)l] n. (Comput.) access to data stored in the memory ▶ The basic purposes of a computer system are information storage and retrieval.

retrieve [rɪˈtriv] v.t. (Comput.) to access data from a computer memory or database ▶ The speed with which data can be retrieved is an important feature of any computer system.

retro- comb. form backward movement

retro-rocket (Mech.) a rocket used to slow down the forward movement of a vehicle by applying reverse thrust ▶ The space shuttle fired its retro-rockets as it came in to land.

rev counter (Mech.) a dial on the dashboard of a vehicle which shows the number of engine revolutions per minute ▶ Changing to a higher gear reduced the revolutions without affecting the speed.

reverse [rɪˈvɜs] v.t. (Mech.) to cause an engine to move in the opposite direction from normal ▶ The pilot reversed the thrust of the engines to slow down the plane after landing.

reverse gear [rɪˌvɜs ˈgɪə] (of a car engine, etc.)

the position of the gears which causes the vehicle to move backwards

revolution¹ [ˌrevəˈluʃ(ə)n] *n.* *(Mech.)* circular movement about a central point ▶ The counter on the fascia board showed an abnormally high number of engine revolutions.

revolution² *n.* *(Mech.)* *(abbr.* **rev)** the turning of a complete circle by a wheel

rheostat [ˈriəstæt] *n.* *(Elec.)* a device by means of which the volume of a sound or the brilliance of an artificial light can be varied by changing the resistance in an electric circuit

rheumatism [ˈruməˌtɪz(ə)m] *n.* *(Med.)* inflammation of the muscles and joints, with painful swelling

rhombus [ˈrɒmbəs] *n.* *(Maths.)* a flat figure with four equal sides but no right angles

Richter [ˈrɪktə] **scale** a scale of 1-10 on which the strength of earth tremors is registered

rifle [ˈraɪf(ə)l] *v.t.* *(Mech.)* to make curved grooves on the inside of a gun barrel so that the shell or bullet will spin and fly straighter

right angle [ˈraɪt ˌæŋg(ə)l] *(Maths.)* an angle formed by two lines meeting perpendicularly

rigor mortis [ˈrɪgə ˈmɔtɪs] *(Med.)* the stiffening of the muscles of a body soon after death ▶ The degree of rigor mortis can help to indicate how long someone has been dead.

ring [rɪŋ] *n.* *(Elec.)* an electrical circuit that links a number of outlets in a building

riser [ˈraɪzə] *n.* *(Build.)* one of the upright parts of a flight of stairs, between the horizontal surfaces ▶ Removing the risers made the staircase much lighter.

RNA *(abbr.)* *(Biol.: Chem.)* ribonucleic acid ▶ RNA and DNA are present in all living cells.

robot [ˈrəʊbɒt] *n.* *(Mech.)* a machine that can do work normally performed by humans ▶ Many stages in the assembly of car parts are now carried out by robots.

rocket [ˈrɒkɪt] *n.* *(Eng.)* a tube-shaped vehicle designed to carry artificial satellites or spacecraft into space and powered by engines which may be discarded at different stages after use ▶ Though the rocket lifted off successfully, the second stage failed to ignite and the mission was aborted.

roentgen [ˈrɜntgən] *n.* *(Phys.: Meas.)* a unit for measuring the strength of X-rays

roll¹ [rəʊl] *v.i.* *(Mech.)* (of motor vehicles, etc.) to move on wheels ▶ The brakes were released and the plane began to roll along the runway.

roll² *v.t.* *(Metall.)* to press heated metal ingots into sheets or bars by passing them between heavy rollers

roll on – roll off (of a ferryboat or cargo ship) one into which and out of which vehicles can drive directly by means of ramps which are lowered on to the quay

roller [ˈrəʊlə] *n.* *(Mech.)* a cylinder of wood, rubber, metal, or other substance for flattening or crushing materials ▶ The chunks of ore are crushed by heavy rollers so that the metal contents can be extracted.

rolling mill [ˈrəʊlɪŋ ˈmɪl] a heavy plant where heated metal ingots are pressed into sheets or bars

rolling stock [ˈrəʊlɪŋ ˌstɒk] the carriages and wagons used on railways ▶ The train was cancelled due to a shortage of rolling stock.

ROM [rɒm] *(abbr.)* *(Comput.)* read only memory, a data-storage device which writes data permanently into the memory of a computer

root [rut] *n.* *(Maths.)* a number which gives another specific number if it is multiplied by itself a given number of times ▶ The square root of 9 is 3 ($3 \times 3 = 9$), the cube root of 27 is 3 ($3 \times 3 \times 3 = 27$), and the fourth root of 81 is 3 ($3 \times 3 \times 3 \times 3 = 81$).

rotary [ˈrəʊtərɪ] *a.* *(Mech.)* (of a movable part) which revolves around a central point ▶ The rotary blades of the powerful mower levelled everything in their path.

rotate [rəʊˈteɪt] *v.i.* *(Mech.)* to revolve around a central point ▶ The Earth rotates around an imaginary axis which joins the

North and South Poles.

rotation [rəʊˈteɪʃ(ə)n] *n. (Mech.)* movement around a central point ▶ The rotation of the Earth about its axis can now be timed precisely.

round¹ [raʊnd] *v.t. (Maths.)* to adjust a figure to the nearest whole number ▶ The actual figures were 97 and 142, but for marking on the graph 97 was rounded up to 100 and 142 was rounded down to 140.

round² *a. (Maths.)* (of a number) whole or full ▶ We ordered 11, but received a round dozen.

round³ *n. (Mech.)* a unit of ammunition for a rifle etc. ▶ He fired six rounds but hit the target only twice.

round figures/round numbers numbers adjusted to the nearest whole number or unit in cases where precise details are not important ▶ In round figures, the population is now about three million.

rudder [ˈrʌdə] *n. (Aer.: Naut.)* a movable upright plane at the back of an aircraft or ship, which controls turning ▶ The rudder was destroyed in the explosion and the ship drifted helplessly on to the rocks.

run [rʌn] *v.i. (Mech.)* (of an engine) to function ▶ Since it was serviced, the engine is running normally again.

runner¹ [ˈrʌnə] *n. (Mech.)* (of a sledge etc.) one of the long, smooth surfaces on which the vehicle runs, in place of wheels

runner² *n. (Bot.)* a stem of a plant which spreads along or below the ground ▶ The wild strawberry plant sent runners all over the flowerbeds.

runway [ˈrʌnweɪ] *n. (Build.)* a long, smooth, roadlike surface along which aircraft run in order to gain speed for take-off or lose speed after landing ▶ The giant plane overshot the tiny runway and came to rest in a field.

rupture [ˈrʌptʃə] *n. (Med.)* a hernia

rust [rʌst] *n. (Chem.)* a reddish encrustation of oxide that forms on iron when exposed for long periods to a moist atmosphere ▶ Streaks of rust covered the hull of the wrecked ship.

rustproof [ˈrʌs(t)ˌpruːf] *a. (Metall.)* (of a metal) treated in some way that prevents it rusting

S

sadism [ˈseɪdɪzm] *n.* (*Psych.*) a mental condition in which sexual pleasure is derived from inflicting physical pain

safety [ˈseɪftɪ] *n.* freedom from danger, injury, or risk

safety belt [ˈseɪftɪ ˌbelt] a belt worn in a motor vehicle or aeroplane to hold the occupant in a seat in case of sudden movement ▶ Since the wearing of safety belts in cars was made compulsory, the death rate in accidents has decreased sharply.

safety catch [ˈseɪftɪ ˌkætʃ] a switch or device on a gun or other mechanism which prevents it from being fired or set in motion accidentally ▶ The safety catch should be kept on whenever the machine is not in use.

safety film [ˈseɪftɪ ˌfɪlm] film made of non-flammable material

safety fuse [ˈseɪftɪ ˌfjuz] a type of fuse that prevents overheating in an electric circuit

safety lamp [ˈseɪftɪ ˌlæmp] a miner's lamp in which the flame is protected from contact with gases in the atmosphere which might cause an explosion ▶ The safety lamp is sometimes called a 'Davey lamp', after its inventor.

safety valve [ˈseɪftɪ ˌvælv] a valve which prevents the pressure of gas etc., in a container from building up to a dangerous level by automatically allowing some of it to escape

saline [ˈseɪlaɪn] *a.* (*Chem.: Med.*) containing a high proportion of a chemical salt ▶ The specimens were preserved in a saline solution. ▶ A saline drip was set up over the bed.

saliva [səˈlaɪvə] *n.* (*Biol.*) a colourless liquid secreted by glands in the mouth to keep it moist, especially for eating ▶ Saliva dripped from the fox's jaws at the sight of the baby rabbits.

salivate [ˈsælɪveɪt] *v.i.* (*Biol.*) to secrete saliva ▶ At the sight of the food, the dog began to salivate and lick its chops.

salmonella [ˌsælməˈnelə] *n.* (*Biol.: Med.*) one of the sorts of bacteria often associated with food poisoning, and the illness caused by these bacteria ▶ The cause of the outbreak of salmonella was badly refrigerated food.

salt [sɔlt] *n.* (*Chem.*) a powder formed by the reaction of a metal with an acid ▶ The most commonly known salt is sodium chloride, which is used in cooking.

salt flat [ˈsɔlt ˌflæt] a flat area on which the evaporation of sea-water has deposited a layer of salt

saltpetre [ˌsɔltˈpitə] *n.* (*Chem.*) a salt used in making gunpowder ▶ The chemical name of saltpetre is potassium nitrate.

satellite[1] [ˈsætəlaɪt] *n.* (*Astron.*) a star that orbits another heavenly body ▶ The Moon is a satellite of the Earth.

satellite[2] *n.* (*Aer.*) a man-made device launched by rocket to go into orbit in space ▶ There are now hundreds of satellites littering space.

saturate [ˈsætʃəreɪt] *v.t.* (*Chem.*) to put as much of a substance into a solution as possible ▶ The tank was first filled with a saturated saline solution.

saturation [ˌsætʃəˈreɪʃ(ə)n] *n.* (*Chem.*) the process or result of saturating a solution

saturation point [ˌsætʃəˈreɪʃ(ə)n ˌpɔɪnt] the point beyond which a solution will not absorb any more of a solid substance ▶ When the solution reached saturation point, the remainder of the salt formed a deposit at the bottom of the tank.

scale¹ [skeɪl] *n.* (*Geog.*) a system of representing distances proportionately on a map, e.g. by making 1 cm on the map represent 1 km on the ground ▶ Because of the small scale of the map, the distance looked much shorter than it actually was.

scale² *n.* (*Maths.*) a table or sequence of units, against which sizes etc. can be measured ▶ The earthquake measured 6·5 on the Richter scale.

scan¹ [skæn] *v.t.* or *n.* (*Med.: Phys.*) to conduct a thorough medical examination by the use of radiography, or the result of such an examination ▶ The scan revealed a tumour in the liver.

scan² *v.t.* or *n.* (*Radio*) to examine a broad area through the use of radar, or the result of such an examination ▶ The radar scan showed an unidentified vessel approaching at speed.

scan³ *v.t.* (*Elec.*) to examine with a laser beam in order to make an image ▶ The document was electronically scanned and transmitted by telephone to its destination.

schist [ʃɪst] *n.* (*Geol.*) a sort of rock that splits easily into layers

schizoid [ˈskɪtsɔɪd] *a.* (*Psych.*) related to or suffering from schizophrenia

schizophrenia [ˌskɪtsə(ʊ)ˈfrɪnɪə] *n.* (*Psych.*) a mental disorder in which a person's mind and feelings are separated, so that they may behave in two quite different ways ▶ The patient was in two minds about the diagnosis of schizophrenia.

science [ˈsaɪəns] *n.* systematic study or knowledge, organized according to logical principles

scientific [ˌsaɪənˈtɪfɪk] *a.* in accordance with the principles of science

scientist [ˈsaɪəntɪst] *n.* a person skilled in the practice of a science

scramble [ˈskræmbl] *v.t.* (*Radio*) to jumble the order of signals in a radio message or telephone conversation so that people not intended to hear will not be able to understand

scrap iron [ˈskræp ˌaɪən] discarded pieces of metal, mostly iron, which can be recycled

screw [skru] *n.* (*Mech.*) a ship's propeller ▶ The wake churned up by the ship's propeller screw was clearly visible long after the vessel had passed.

scupper [ˈskʌpə] *n.* (*Naut.*) an opening in the side of a ship's deck to let the water run off ▶ A huge wave broke over the decks and ran off through the scuppers.

scuttle¹ [ˈskʌtl] *n.* (*Naut.*) an opening with a removable lid in the side of a ship

scuttle² *v.t.* (*Naut.*) to sink a ship by opening the scuttles and letting the water in ▶ Rather than surrender his ship, the captain scuttled it in the open sea.

season [ˈsiz(ə)n] *v.t.* (*Build.*) (of timber) to harden by drying gradually ▶ Wood that has not been properly seasoned will very soon crack or split.

sec. (*abbr.*) second

second¹ [ˈsek(ə)nd] *n.* (*Phys.: Maths.*) one sixtieth of a minute in time

second² *n.* (*Geog.*) one-sixtieth of a minute of longitude or latitude, which is one-sixtieth of a degree ▶ The map-reference we were given was 10 degrees (10°), 12 minutes (12′) sixteen seconds (16″) N, and 5° 10′ 40″ W.

secondary [ˈsek(ə)nd(ə)rɪ] *a.* not of primary level or importance

secondary cell an electric cell that can be recharged or used to store energy

secondary colour a colour formed by mixing two primary colours ▶ A mixture of the two primary colours yellow and blue will give the secondary colour green.

secrete [sɪˈkrit] *v.t.* (*Biol.*) to give out a fluid substance ▶ Saliva is secreted from glands in the mouth.

secretion [sɪˈkriʃ(ə)n] *n.* (*Biol.*) the process or result of secreting a fluid ▶ The secretion from cuts in the bark of trees is collected and processed as rubber.

section¹ [ˈsekʃ(ə)n] *n.* (*Maths.*) the shape of a figure obtained when a solid body is sliced through

section² *n.* (*Biol.*) a thin slice of tissue from a bodily organ or other organic matter ▶

The botanist sliced the strange growth into sections and examined them under a microscope.

sector ['sektə] *n. (Maths.)* the area of a circle between two lines drawn from the centre to the circumference

sedate [sɪ'deɪt] *v.t. (Med.)* to calm a patient down by administering a drug ▶ The hysterical woman was sedated and eventually became quiet.

sedation [sɪ'deɪʃ(ə)n] *n. (Med.)* the process or result of sedating ▶ The police were told that the victim was under sedation and could not yet be interviewed.

sedative ['sedətɪv] *n. (Med.)* a drug that has a calming effect ▶ The doctor gave her a mild sedative to help her to overcome the shock.

sediment ['sedɪ'mənt] *n. (Chem.: Geol.)* small particles deposited at the bottom of a body of liquid ▶ Analysis of sediment from the bottom of the lake showed traces of toxic waste.

sedimentary [ˌsedɪ'ment(ə)rɪ] *a. (Chem.: Geol.)* composed of sediment

sedimentary rock a layer of rock formed from sediment deposited many years earlier

sedimentation [ˌsedɪmen'teɪʃ(ə)n] *n. (Chem.)* the process or result of depositing sediment ▶ Over a number of years the process of sedimentation gradually blocked the outlet from the tank.

seep [sip] *v.i. (Phys.)* (of a liquid) to leak through a crack or hole ▶ Oil seeping from a crack in the fuel pipe caused a violent explosion.

seepage ['sipidʒ] *n. (Phys.)* the process or result of seeping ▶ The explosion was attributed to seepage of oil through a crack in the fuel pipe.

segment ['segmənt] *n. (Maths.)* the area between a line drawn through a circle and the circumference of the circle

seismic ['saɪzmɪk] *a. (Geol.)* having to do with an earth tremor

seismograph ['saɪzməˌgrɑf] *n. (Geol.)* an instrument for measuring the strength of earthquakes

seismologist [ˌsaɪz'mɒlədʒɪst] *n. (Geol.)* an expert in the science of seismology ▶ Seismologists cannot yet predict with certainty where and when earthquakes are likely to take place.

seismology [ˌsaɪz'mɒlədʒɪ] *n. (Geol.)* the branch of science concerned with the study of earth tremors (earthquakes)

seize (up) [ˌsiz 'ʌp] *v.i. (Mech.)* (of an engine) to cease functioning as the result of overheating ▶ Because of a leak in the radiator, the engine overheated and seized up.

semi- *comb. form* half, partly

semiconductor [ˌsemɪkən'dʌktə] *n. (Elec.)* a substance which has a reduced resistance to the passage of an electrical current when heated ▶ The discovery of semiconductors was an important step in the development of transistor radios.

semiconscious [ˌsemɪ'kɒnʃəs] *a. (Med.)* half or only partly conscious

send [send] *v.t. (Radio)* to transmit a signal by radio ▶ The radio operator managed to send out a call for assistance before the ship went down.

sensitive ['sensətɪv] *a. (Mech.: Photo.)* (of a piece of apparatus or film) capable of detecting slight changes of movement, brilliance of light, pressure of electrical current etc. ▶ The thermometer was so sensitive that it detected the body-heat of anyone who entered the laboratory ▶ A very sensitive film was used to photograph the interior of the cave.

sensitivity [ˌsensə'tɪvətɪ] *n.* the quality or degree of being sensitive

sensor ['sensə] *n.* a piece of apparatus that receives a signal and responds to it ▶ A common form of sensor is the electric eye in the doorway of a lift or elevator.

sepsis ['sepsɪs] *n. (Med.)* poisoning of part of the body by the action of bacteria, often visible through the formation of pus ▶ Wounds which are not properly cleaned may develop sepsis and become more serious.

septic ['septɪk] *a. (Med.)* infected with sepsis ▶ The septic wound was treated with an

antibiotic paste.
septic tank a tank in which waste matter from a lavatory etc., is collected and treated chemically through the action of bacteria ▶ Since the house was too remote to be connected to the main sewage system, it was equipped with a septic tank.
sequence ['siːkwəns] *n.* (*Maths.*) a set of numbers with a fixed order and relationship ▶ 2, 4, 6, 8 is a sequence in which items are progressively increased by 2 (2+2=4, +2=6, +2=8).
series[1] ['sɪəriːz] *n.* (*Maths.*) the set of numbers in a sequence ▶ The series 2-4-6-8 is an example of an arithmetical progression.
series[2] *n.* (*Elec.*) a number of electrical components connected in such a way that the current flows through each of them in turn ▶ The circuit was made up of a variable number of components, connected in series.
serum ['sɪərəm] *n.* (*Biol.: Med.*) the watery fluid in blood, which can be extracted and injected into another body to combat infection ▶ Supplies of serum were flown by helicopter to the scene of the disaster.
servomechanism ['sɜːvəʊˌmekənɪz(ə)m] *n.* (*Mech.*) a mechanism that supplies power to a machine by converting a small force into a larger one ▶ One of the most primitive but effective forms of servomechanism is the pulley.
set[1] [set] *n.* (*Biol.: Hort.*) a cutting, bulb, etc., that can be used to grow a new plant
set[2] *n.* (*Maths.*) any collection of items with a common factor
sewage ['sjuːɪdʒ] *n.* (*Eng.*) domestic or industrial waste matter which is carried away in drains for treatment and disposal ▶ The city's sewage system was built a hundred years ago and is no longer adequate.
sewage farm ['sjuːɪdʒ ˌfɑːm] a place where sewage is treated ▶ The smell from the sewage farm was intolerable when the wind came from that direction.
sewer ['sjuːə] *n.* (*Build.*) a drain that carries away sewage ▶ Hidden beneath the streets of every city is a vast network of sewers which few people ever see.
sewerage ['sjuːərɪdʒ] *n.* (*Build.*) the system for removing waste matter through sewers ▶ The sewerage in some of the older cities is in danger of breaking down.
sextant ['sekst(ə)nt] *n.* (*Geog.: Naut.*) an instrument used for navigating by reference to stars
shaft[1] [ʃɑːft] *n.* (*Mech.*) a rod around which a wheel or belt revolves ▶ When the lubricating system on the central shaft broke down, the machinery overheated and seized up.
shaft[2] *n.* (*Mining*) the main channel through which miners reach the coalface and through which coal etc. is brought above ground ▶ The miners descend into the shaft in an open mesh lift called a cage.
shaft[3] *n.* (*Eng.*) a hollow metal tube through which air etc. is distributed ▶ The ventilator shaft led from the basement right up to the roof.
shear [ʃɪə] *v.i.* (*Mech.*) (of a metal pin, bolt, etc) to snap as the result of pressure from one side ▶ When the truck ran over a boulder, the back axle sheared and the vehicle was immobilized.
sheath[1] [ʃiːθ] *n.* (*Mech.*) a sleeve or covering to hold or protect something ▶ The blade was protected by a thick leather sheath.
sheath[2] *n.* (*Med.*) a condom
shock[1] [ʃɒk] *n.* (*Elec.: Med.*) the sudden and violent jolt experienced when the body comes in contact with an electric current ▶ The electric leads had not been properly connected and gave him a violent shock when he pulled the plug out of the socket.
shock[2] *n.* (*Med.*) the acutely weakened state of the body after a person has suffered a sudden injury or fright ▶ Though not injured in the fall, he was taken to hospital and treated for shock.
shock absorber ['ʃɒk əbˌsɔːbə] a mechanism on the wheels of a vehicle which lessens the effect of going over a bump or of landing heavily
shock wave ['ʃɒk ˌweɪv] a sudden blast of air following an explosion ▶ The explosion

shoot¹ [ʃut] *v.t.* (*Mech.*) (of a bolt) to close firmly ▶ She slammed the door and shot the bolt so that no one could enter.

shoot² *v.t.* (*Naut.*) to establish the position of a ship, etc., by pointing a sextant at a star

shoot³ *v.t.* (*Photo.*) to photograph something, especially when it is moving ▶ We used several reels of film shooting the animals in the safari park.

shoot⁴ *n.* (*Bot.*) the new growth of a plant ▶ The panda feeds only on certain bamboo shoots.

short [ʃɔt] *v.i.* (*Elec.*) (of an electric circuit) to be broken because the current has taken too short a route ▶ The crossed wires caused the circuit to short and started a fire.

short circuit [ˌʃɔt ˈsɜkɪt] the breakdown that occurs in an electric circuit when the current takes too short a route

short-sighted [ˌʃɔtˈsaɪtɪd] (of a person) unable to see or read anything well unless it is close to the eyes.

short-wave [ˈʃɔtˌweɪv] (of a radio signal) transmitted or received on waves of less than 60 metres ▶ The crippled yacht contacted the coastguard by short-wave radio.

shot¹ [ʃɒt] *n.* (*Aer.*) the launch of a rocket carrying a satellite or probe ▶ The success of the next shot from Cape Canaveral was vital to the continuation of the space programme.

shot² *n.* (*Photo.*) a photograph, especially one taken quickly of a moving object ▶ An amateur photographer in the crowd took some spectacular shots of the charging lion.

shrink [ʃrɪŋk] *v.i.* and *v.t.* to make or become smaller as the result of some outside influence ▶ The material was artificially shrunk before the garments were made.

shrink wrap to wrap an object in an airtight plastic covering especially shrunk to fit it

shunt¹ [ʃʌnt] *n.* (*Elec.*) a device for providing an alternative route for an electric current

shunt² *v.t.* (*Elec.*) to divert a specific part of an electric current by the use of a low-resistance device, or shunt, in the circuit

shutter [ˈʃʌtə] *n.* (*Mech.*) a movable part, used to cover or uncover an opening ▶ Opening the shutter of a camera exposes the film to light.

shuttle [ˈʃʌtl] *n.* (*Mech.*) a part of a mechanism that slides backwards and forwards ▶ Threading the cotton through the shuttle is the first step in operating a sewing machine.

sieve [sɪv] *n.* (*Mech.*) an instrument for separating small pieces of matter from larger pieces, or solid particles from a liquid, by pouring the mixture through a wire mesh

sight [saɪt] *n.* (*Mech.*) (on a rifle) a notched metal plate which enables the user to take accurate aim ▶ By using a telescopic sight you can see the target much more clearly.

signal [sɪgn(ə)l] *n.* (*Radio*) a pattern of sounds or light which can be recognized and understood ▶ A faint distress signal was heard, and eventually the missing aircraft was found.

silica [ˈsɪlɪkə] *n.* (*Geol.: Chem.*) a natural substance, usually found in the form of sand, which is used in making glass

silicate [ˈsɪlɪkət] *n.* (*Geol.*) a group of natural substances, from which many of the Earth's minerals are made up

silicon [ˈsɪlɪkən] *n.* (*Chem.*) a non-metallic element found naturally in a number of compound forms ▶ Silicon is used in making such items as radio transmitters.

silicon chip [ˌsɪlɪkən ˈtʃɪp] a silicon microchip used for making integrated circuits for computerized equipment

silicone [ˈsɪlɪkəʊn] *n.* (*Chem.*) a group of synthetic compounds which are not affected by changes of temperature and which resist the effects of humidity ▶ The use of silicone paints makes frequent redecoration unnecessary.

silicosis [ˌsɪlɪˈkəʊsɪs] *n.* (*Med.*) a disease of the lungs caused by the inhalation of silica dust ▶ Many coal miners died of pneumonia as a consequence of silicosis of the lungs caused by inhaling dust.

simple ['sɪmpl] *a.* in its basic form, not combined with anything else

simple fraction a fraction in which both numerator and denominator are simple numbers ▶ Both $\frac{1}{2}$ and $\frac{1}{3}$ are simple fractions, sometimes called 'vulgar' fractions.

simple fracture a breakage of a bone not involving any other tissue ▶ The break in the cyclist's leg was a simple fracture, so there was no displacement and no bleeding.

simple machine one of the elementary devices, such as a wheel or pulley, from which machines are built ▶ Primitive man was adept in the use of such simple machines as the lever and the wheel.

simplify ['sɪmplɪfaɪ] *v.t.* (*Maths.*) to reduce (a fraction etc.) to its basic parts

simulate ['sɪmjʊleɪt] *v.t.* (*Comput.: Eng.*) to produce a likeness of something in order to study it and make predictions about it ▶ The appearance of the new car model was simulated on the computer screen.

simulation [ˌsɪmjʊ'leɪʃ(ə)n] *n.* (*Comput.*) the process or result of simulating something ▶ The result of the simulation of the new model was the introduction of some significant modifications.

simulator ['sɪmjʊleɪtə] *n.* (*Eng.*) a machine that simulates the effects of a specific situation or experience ▶ The crew of the space shuttle had been familiarised in a simulator with the condition of weightlessness.

simultaneous [ˌsɪməl'teɪnɪəs] *a.* proceeding at the same time as something else

simultaneous equation one of a number of equations having more than one variable, each of these variables having the same value in all of the equations

sink [sɪŋk] *v.t.* (*Mining*) to dig a mine or shaft ▶ A series of shafts were sunk along the floor of the valley, and rich seams of coal were tapped.

skid [skɪd] *v.i.* (*Mech.*) (of a wheel) to slip sideways because of failure to gain proper purchase on a slippery surface ▶ The lorry skidded on the icy surface and went into the ditch at the side of the road.

skip [skɪp] *n.* (*Build.*) a large, portable metal container for transporting rubble ▶ A skip was hired for several days while the wall was demolished.

skirt [skɜt] *n.* (*Mech.*) a protective covering over part of a lathe to prevent injury to the operator from flying fragments of metal etc.

slack [slæk] *n.* (*Mech.*) (of a hawser, etc.) that part which is not being stretched ▶ The winch took up the slack and the hawser tightened around the stanchion.

slake [sleɪk] *v.t.* (*Chem.*) (of lime) to mix with water in order to form a chemical compound, calcium hydroxide

slave [sleɪv] *n.* (*Mech.*) a part of a machine that depends for its function on another part ▶ The cassettes were copied in batches of five by a machine on which a dubbing master drove five slaves.

sleeve [sliv] *n.* (*Mech.*) a tubular machine part which fits over another part in order to make it bigger ▶ A metal sleeve was fitted around the piston to ensure a tighter fit inside the cylinder.

slick [slɪk] *n.* (*Ecol.*) a mass of oil floating on the sea ▶ Ecologists were concerned that the oil slick from the wrecked tanker was threatening sea birds and marine animals.

slide [slaɪd] *n.* (*Biol.*) a thin piece of glass, on which sections are placed for examination under a microscope ▶ The stem of the plant was sliced into sections and mounted on slides for examination.

slot[1] [slɒt] *n.* (*Mech.*) a narrow, straight-sided opening, through which objects may be inserted, such as coins into a vending machine ▶ The driver inserted a pound coin in the slot and received a car-park ticket for one hour.

slot[2] *n.* a place in a tightly organized schedule ▶ The incoming plane was late arriving, so the outgoing passengers had to wait several hours until a new departure slot could be found.

sluice [slus] *n.* (*Eng.*) a channel through which the flow of water can be controlled by means of a movable gate ▶ When the water in the river rose to a dangerous level, some of it was diverted through a sluice around the weir.

smelt [smelt] *v.t.* *(Metall.)* to heat ore to a high temperature in order to extract its metal content ▶ The Iron Age is so called because it was then that Man first learned how to smelt iron.

socket [ˈsɒkɪt] *n.* *(Mech.: Elec.)* a hollow slot or hole in a mechanism, into which another part of the mechanism fits or a device into which an electric plug or lamp may be fitted in order to make a connection ▶ Movement was facilitated by a ball and socket joint. ▶ She plugged the radio into a socket in the wall.

sodium [ˈsəʊdɪəm] *n.* *(Chem.)* a silver-coloured, non-metallic element found in combination with various other elements

sodium chloride [ˌsəʊdɪəm ˈklɔraɪd] common salt, as used in cooking

soft [sɒft] *a.* malleable, pliable, easily changed

soft drug a drug, such as cannabis which is not habit-forming

soft water water with a low content of the minerals which stop soapsuds from forming ▶ Rainwater is soft water.

software [ˈsɒf(t)weə] *n.* *(Comput.)* the programs contained in disks etc. which are used by computers ▶ When the new computer system was installed, the old software was found to be incompatible.

solar [ˈsəʊlə] *a.* *(Phys.: Astron.)* having to do with the Sun or with sunlight

solar cell [ˌsəʊlə ˈsel] a device for converting sunlight into energy, e.g. for heating water

solar panel [ˌsəʊlə ˈpæn(ə)l] a panel erected on the roof of a building, with solar cells for heating water or producing electricity ▶ All new buildings in the Greek islands are equipped with solar panels.

solar system [ˌsəʊlə ˈsɪstəm] a sun and the planets that orbit it ▶ It is now known that the solar system to which Earth belongs is only one of many.

solar year [ˌsəʊlə ˈjɪə] the length of time it takes for the Earth to complete an orbit of the Sun ▶ The length of the solar year is 365 days, 5 hours, and 49 minutes.

solder[1] [ˈsɒʊldə] *n.* *(Metall.)* a soft compound of tin and lead, which melts and hardens quickly ▶ Solder is used by electricians for joining wires, and by plumbers for sealing pipe-joints.

solder[2] *v.t.* *(Metall.)* to join pieces of metal with solder

soldering iron [ˈsɒʊld(ə)rɪŋ ˌaɪən] an implement with an electrically heated blade for joining connections in circuits etc.

solenoid [ˈsəʊlənɔɪd] *n.* *(Elec.)* a coil of wire made to move by the magnetic field created by an electrical current, in order to operate a switch ▶ A solenoid is the mechanism that operates the starter motor of a petrol engine.

solid [ˈsɒlɪd] *n.* *(Chem.)* a substance that does not flow like a liquid or a gas ▶ At low temperatures many liquids become solids, but at very high temperatures solids may first become liquid and then vapour.

solid fuel[1] [ˌsɒlɪd ˈfjʊəl] coal, coke, wood etc. for burning in fires, furnaces or stoves

solid fuel[2] highly volatile fuel for the engines of rockets etc. which is stored at low temperatures in solid form

solid geometry [ˌsɒlɪd dʒɪˈɒmətrɪ] the branch of geometry concerned with solid figures, as opposed to plane or flat figures

solid state [ˌsɒlɪd ˈsteɪt] (of electronic components) comprising semiconductors etc. as in transistor radios

solstice [ˈsɒlstɪs] *n.* *(Astron.)* one of the two dates in the year when night and day are of the same length ▶ The summer solstice is 21 June, and the winter solstice is 22 December.

solubility [ˌsɒljʊˈbɪlətɪ] *n.* *(Chem.)* the degree to which a substance will dissolve in a solvent ▶ Experiments were conducted to establish the solubility of the newly isolated element in a variety of solvents.

soluble [ˈsɒljʊb(ə)l] *a.* *(Chem.)* capable of dissolving ▶ Soluble aspirins are easier to take.

solution[1] [səˈluːʃ(ə)n] *n.* *(Chem.)* a liquid mixture obtained by dissolving one substance in another ▶ The metal filings were immersed in a solution of concentrated sulphuric acid.

solution² *n. (Maths.)* the answer to a mathematical problem

solvent ['sɒlv(ə)nt] *n. (Chem.)* a liquid in which another substance may dissolve ▶ Several possible solvents were tried, but the plastic material would not dissolve in any of them.

sonar ['səʊnɑ] *(abbr.) n. (Radio.: Naut.)* sound navigation ranging, a system of measuring depth below the surface of water by sending out a radio signal and timing the echo which is reflected off an object struck by the signal

sonic ['sɒnɪk] *a. (Phys.)* having to do with sound or the making of sound

sonic barrier [ˌsɒnɪk ˈbærɪə] the resistance encountered by aircraft in exceeding the speed of sound ▶ Breaking through the sonic barrier opened the way to supersonic flight. ▶ The sonic barrier may also be called the 'sound barrier'.

sonic boom [ˌsɒnɪk ˈbuːm] a bang produced when an aircraft, etc., breaks through the sonic barrier ▶ Producing sonic booms is prohibited over land.

sound¹ [saʊnd] *v.t. (Naut.)* to measure the depth of the sea by the use of a calibrated line or rod

sound² *n. (Phys.)* the sensation produced by the organs of hearing, perceived via the ears

sound barrier [saʊnd ˌbærɪə] another name for sonic barrier

sour [saʊə] *a. (Chem.)* (of a taste) sharp or acidic ▶ Vinegar is used in some kinds of cooking to give a sour flavour to a dish that might otherwise seem too sweet.

space [speɪs] *n. (Phys.: Astron.)* the universe beyond the Earth's atmosphere ▶ The development of rocket technology has enabled us to achieve the age-old dream of conquering space.

space age ['speɪs ˌeɪdʒ] the period in which it has been possible to travel in space

space capsule ['speɪs ˌkæpsjuːl] a vehicle sent into space from Earth and returning again to Earth

space man/woman ['speɪs ˌmæn/ˌwʊmən] a man/woman who travels in space ▶ The first space woman was the Russian, Valentina Tereshkova.

space platform ['speɪs ˌplætfɔm] a spacecraft designed to remain for a lengthy period in space and to be used as a base for the dispatch of other spacecraft

space probe ['speɪs ˌprəʊb] a space satellite or other vehicle sent into space to transmit scientific data back to Earth

space shuttle ['speɪs ˌʃʌtl] a spacecraft that is launched into space from another spacecraft and lands on Earth like an aircraft ▶ A space shuttle can be used many times.

space station ['speɪs ˌsteɪʃ(ə)n] another term for space platform

space suit ['speɪs ˌsjuːt] a sealed and pressurized suit worn on space missions to provide its own atmosphere during space walks

space walk ['speɪs ˌwɔːk] work done in space outside a spacecraft by a space man or woman in a space suit

spacecraft ['speɪs ˌkrɑːft] *n. (Aer.)* a vehicle sent into space, usually to orbit Earth to observe other bodies and relay information to Earth

spaceship ['speɪsˌʃɪp] *n. (Aer.)* a manned spacecraft

spark [spɑːk] *n. (Elec.: Mech.)* a controlled electrical discharge which ignites the fuel in a petrol engine ▶ If the points are corroded, they will not produce a spark.

spark chamber ['spɑːk ˌtʃeɪmbə] the chamber in which a controlled spark is produced

spark gap ['spɑːk ˌgæp] the distance between the two electrodes across which a spark must jump

spark plug ['spɑːk ˌplʌg] a device fixed into the cylinder of a petrol engine as a means of producing an electric spark to ignite the fuel

species ['spiːʃiːz] *n. (Biol.)* a taxonomic subdivision of a genus, into which plants etc.

capable of breeding together are categorized ▶ Environmentalists are worried about the increasing number of endangered species as a result of industrial pollution.

specific [spəˈsɪfɪk] *a. (Chem.: Phys.)* characteristic of a given substance ▶ The specific properties of the newly developed alloy are now being extensively tested.

specific gravity [spəˌsɪfɪk ˈgrævɪtɪ] the ratio of the density of any substance compared with that of water

specific heat [spəˌsɪfɪk ˈhit] the amount of energy required to heat a given quantity of a substance by a given amount in given circumstances

specific volume [spəˌsɪfɪk ˈvɒljum] the ratio of a volume of a substance to its weight

specification [ˌspesɪfɪˈkeɪʃ(ə)n] *n.* a description of work to be done, materials required, size etc.

specify [ˈspesɪfaɪ] *v.t.* to make a detailed or precise specification

specimen [ˈspesɪmən] *n. (Biol.)* an example or sample intended to typify a whole ▶ Specimens of the rocks found on the Moon were brought back to Earth for analysis. ▶ A specimen of the crash driver's urine was sent to the police laboratory for analysis.

spectrograph [ˈspektrə(ʊ)ˌgrɑf] *n. (Phys.)* a device for producing or recording a spectrum

spectroscope [ˈspektrə(ʊ)ˌskəʊp] *n. (Phys.)* a device for observing a spectrum

spectrum [ˈspektrəm] *n. (Phys.)* the band of colours into which white light is split when it passes through a prism ▶ The constituent colours of white light which form a spectrum are violet, indigo, blue, green, yellow, orange and red.

speed [spid] *n. (Photo.)* the relative sensitivity of a film to light, measured in DIN ▶ The speed of the film was not adequate to make good shots of the action.

speleology [ˌspelɪˈɒlədʒɪ] *n. (Geol.)* the science of investigating the geology of underground caves ▶ Speleology can be an extremely dangerous occupation, since many caves are flooded for much of the time.

sphere [sfɪə] *n. (Maths.)* a solid figure, like a ball, on which all points on the surface are the same distance from the centre ▶ We now know that the Earth is not a perfect sphere.

spherical [ˈsferɪk(ə)l] *a. (Maths.)* in the shape of a sphere

spindle [ˈspɪnd(ə)l] *n. (Mech.)* a rod or similar machine-part that revolves ▶ The flywheel was mounted on a delicate spindle of finest steel.

spinal column the backbone, an interconnected string of vertebrae

spinal cord the nerve fibres which run along the inside of the spinal column and are connected with the brain

spine [spaɪn] *n. (Biol.)* backbone

spiral [ˈspaɪr(ə)l] *a. (Maths.)* rising or dropping by revolving around a central point ▶ A magnificent spiral staircase rose from the centre of the theatre lobby.

spirit [ˈspɪrɪt] *n. (Chem.)* a distilled alcoholic liquid, such as brandy, gin or whisky

spirit level [ˈspɪrɪt ˌlev(ə)l] an instrument for checking whether surfaces are horizontal by reference to the position of an air bubble observed through the window in a sealed tube containing a spirit

spoil [spɔɪl] *n. (Mining)* unwanted material after coal, etc., has been extracted ▶ The countryside around the pit-head was marred by huge tips of spoil from the mine.

sport [spɔt] *n. (Biol.)* a plant or animal that differs strikingly in some way from the rest of its species ▶ Atomic fallout produced a number of sports in the local flora and fauna.

spring [sprɪŋ] *n. (Mech.)* a coil or strip of metal which resumes its original length and shape after being stretched or depressed ▶ Strong springs, called shock absorbers, deaden the effect when a vehicle runs over bumps in the road.

spring balance a device for weighing objects by

suspending them from a spring, the extension of which can be measured against a calibrated scale

sprocket ['sprɒkɪt] *n. (Mech.)* one of a set of teeth on the rim of a wheel, that turns by engaging a chain

sprocket wheel ['sprɒkɪt ˌwil] a wheel which is driven round by projecting teeth that engage with a chain or other moving mechanism ▶ The movement of the sprocket wheel in a film cassette ensures that the same frame cannot be exposed twice.

Sputnik ['spʊtnɪk] *n. (Aer.)* any one of a series of Russian space satellites

square^1 [skweə] *n. (Maths.)* a right-angled plane figure with four equal sides

square^2 *n. (Maths.)* the product obtained by multiplying a number by itself ▶ The square of 2 (2^2) is 4, and of 4 (4^2) is 16.

square root [ˌskweə 'rut] the number which gives another specified number when multiplied by itself ▶ If the square root of 4 is 2, and of 16 is 4, what is the square root of 100?

stability^1 [stə'bɪlətɪ] *n. (Mech.)* the quality that enables mechanical or electrical systems to return to a stable state after being disturbed ▶ The stability of the new car in cornering at speed is a major sales feature.

stability^2 *n. (Chem.)* (of a substance) the quality of not readily combining with other substances to form compounds ▶ Scientists are unsure of its stability at high temperatures.

stabilize ['steɪbəlaɪz] *v.i.* and *v.t.* (Mech.) to return to a stable position

stabilizer [steɪbəlaɪzə] *n. (Aer.: Naut.)* a device that gives extra stability to an aircraft or ship in conditions of turbulence ▶ The addition of stabilizers to cross-channel ferries has made them safer for transporting heavy vehicles.

stable^1 ['steɪb(ə)l] *a. (Mech.)* steady and resistant to violent changes of position ▶ Despite the heavy seas, the little boat remained stable.

stable^2 *a. (Chem.)* (of a substance) resistant to chemical change ▶ At room temperature, the substance was normally stable, but when heated it lost its stability.

stage^1 [steɪdʒ] *n. (Elec.)* part of an electrical circuit consisting of several relays

stage^2 *n. (Eng.)* one of the parts of a multi-stage rocket which ignites in turn to take the rocket further along its trajectory ▶ As the second stage ignited, the rocket disappeared into the clouds.

stalactite ['stæləktaɪt] *n. (Geol.)* a deposit of carbonate of lime which hangs like an icicle from the roof of a cave, formed where water drops from the roof ▶ The giant stalactites took hundreds of thousands of years to form.

stalagmite ['stæləgmaɪt] *n. (Geol.)* a deposit of carbonate of lime which rises in a thin column from the floor of a cave ▶ Stalagmites are formed from the drips from stalactites and may sometimes join with them to form complete pillars from floor to roof.

stall [stɔl] *v.i. (Mech.)* (of an engine) to cut out, due to mechanical failure or mishandling ▶ Because of the low-grade fuel, the engine stalled whenever the car stopped.

stanchion ['stanʃ(ə)n] *n. (Eng.)* a solid pillar or prop supporting a structure

stand [stænd] *n. (Mech.)* a frame or platform to which something is fixed in order to remain upright ▶ The test-tube was clamped into a stand and gently heated over a bunsen burner.

standard ['stændəd] *n.* an agreed level of size, speed, performance, length etc. against which other examples may be measured ▶ The workers agreed to be paid the standard wages, though these were not very high.

standard time [ˌstændəd 'taɪm] the accepted method of reckoning time by reference to an agreed meridian ▶ Standard time is measured with reference to the Greenwich meridian.

starboard ['stɑbəd] *n. (Naut.: Aer.)* the right-hand side of a ship, aircraft etc. ▶ The green navigation light told us that we were passing the starboard side of the other plane.

starch [stɑtʃ] *n.* (*Biol.*) a tasteless, white substance found in most plants and forming an important part of the human diet

start [stɑt] *v.t.* (*Mech.*) to bring a motor into action, usually by switching on the ignition ▶ On damp and cold mornings the engine can be very difficult to start.

starter (motor) *n.* (*Mech.*) a device which starts a petrol engine by igniting the fuel through an electric spark ▶ The engine was immobilised because of a faulty starter.

static [ˈstætɪk] *a.* immobile, unchanging

static electricity electrical energy caused by friction, and stationary, not moving in a current ▶ Radio reception is sometimes subject to interference from static electricity. ▶ If you brush a rubber balloon against a wall it will stick to it through the static electricity generated by the friction.

static tube a tube used to measure the pressure at a fixed point in a moving fluid ▶ The flow of water through the sluice is regulated by the use of a static tube at the entrance.

statics [ˈstætɪks] *n.* the branch of mechanics concerned with creating a state of equilibrium

station [ˈsteɪʃ(ə)n] *n.* (*Comput.*) one of the positions in a network where a terminal is installed

stay [steɪ] *n.* (*Mech.*) a strut, usually of metal, which holds part of a machine or structure in place ▶ The wings of the tiny plane were supported by flimsy wooden stays.

stellar [ˈstelə] *a.* (*Astron.*) having to do with the stars

stellate [ˈsteleɪt] *a.* (*Maths.*) having the shape of a star

stereo- *comb. form* three-dimensional

stereograph [ˈsterɪə(ʊ)ˌgrɑf] *n.* (*Phys.*) a set of two almost identical photographs which, when viewed through special glasses, give a three-dimensional impression

stereophonic [ˌsterɪə(ʊ)ˈfɒnɪk] *usually* **stereo** *a.* (*Phys.*) made by a system of recording or reproducing sound by the use of two microphones and two speakers ▶ Stereo recordings have now replaced the old mono versions.

stereoscope [ˈsterɪəˌskəʊp] *n.* (*Phys.*) a device for viewing sets of nearly identical photographs to obtain a three-dimensional effect

sterile¹ [ˈsteraɪl] *a.* (*Biol.*) unable to reproduce itself ▶ The radiation from the nuclear plant made much of the wildlife of the neighbourhood sterile.

sterile² *a.* (*Med.*) free from living organisms or germs ▶ Surgical operations must be performed with sterile instruments in order to avoid accidental infection.

sterility [stəˈrɪlətɪ] *n.* (*Med.*) the state of being sterile ▶ An abnormally high rate of sterility was discovered in young people working in the nuclear plant.

sterilization [ˌsterəlaɪˌzeɪʃ(ə)n] *n.* (*Biol.: Med.*) the act of rendering sterile ▶ Attempts to control the growth of population by techniques of sterilisation have not been successful.

steroid [ˈstɪərɔɪd] *n.* (*Chem.: Med.*) one of a group of organic compounds with a specific structure which are used in the treatment of certain illnesses, notably cancer ▶ The use of body-building steroids by Olympic athletes is totally forbidden.

stethoscope [ˈsteθəˌskəʊp] *n.* (*Med.*) an instrument used for listening to body noises, such as heartbeats and breathing ▶ Listening through the stethoscope revealed an irregularity in the patient's heartbeat.

stigma [ˈstɪgmə] *pl.* **stigmata** stɪgˈmɑtə] *n.* (*Med.*) a mark on the skin as the result of a disease

stimulant [ˈstɪmjʊlənt] *n.* (*Med.*) a drug which heightens the activity of a body or part of a body ▶ Alcohol is often taken as a stimulant, but its ultimate effect is to depress.

stimulate [ˈstɪmjʊleɪt] *v.t.* (*Med.*) to excite a nerve, etc., by the use of a stimulant

stimulus [ˈstɪmjʊləs] *n.* (*Biol.: Psych.*) something which excites a living organism or prompts action

stirrup pump [ˈstɪrəp ˌpʌmp] a hand-operated

device for pumping water out of a bucket to fight fires

stomach pump [ˈstʌmək ˌpʌmp] a suction pump used to remove the contents of the stomach ▶ The poison that the patient had swallowed was sucked out with a stomach pump.

stopcock [ˈstɒpˌkɒk] n. (Mech.) a valve used to control the flow of water through a pipe

storage [ˈstɔrɪdʒ] n. (Comput.) the process or result of loading data into a computer for retrieval or processing according to the program in use

strain [streɪn] v.t. to remove pieces of matter from a liquid by pouring it through a sieve

stratification [ˌstrætɪfɪˈkeɪʃ(ə)n] n. (Geol.) the natural arrangement of sedimentary rock in strata, or layers, which indicate their relative age ▶ The stratification of the local rocks was clearly visible in the sides of the quarry.

stratosphere [ˈstrætə(ʊ)ˌsfɪə] n. (Phys.) the layer of the Earth's atmosphere between the troposphere and the mesosphere

stratum [ˈstrɑtəm] pl. **strata** n. (Biol.: Geol.) a layer of tissue or rock

stress [stres] v.t. (Build.) to exert pressure on something in order to test its strength ▶ The materials employed in constructing the new tower-block were stressed before use.

strike [straɪk] v.t. (Mining) (of a mineral etc.) to discover after a process of prospecting ▶ After months of fruitless search they eventually struck oil.

stroke [strəʊk] n. (Med.) a sudden interruption of the flow of blood to the brain, resulting in weakness or paralysis ▶ On the day of her retirement she suffered a severe stroke which left her incapable of speech.

structure¹ [ˈstrʌkfʃə] n. (Chem.) the arrangement of atoms within a molecule of a given substance ▶ The chemical structure of the new element has not yet been established.

structure² n. (Build.: Eng.) the framework on which something is constructed ▶ Excavations have revealed details of the structure of mediaeval buildings.

strut [strʌt] n. (Eng.) a rod or strip, usually of metal, supporting part of a structure ▶ The wings were supported by struts made of a light aluminium alloy.

stucco [ˈstʌkəʊ] n. (Build.) a plaster mixture used to coat the surface of a wall ▶ The stucco surface of the farmhouse building had been rotted away by the damp.

styptic [ˈstɪptɪk] a. (Med.) containing a drug which stops bleeding by making the exposed ends of blood vessels contract

styptic pencil a stick of material containing a styptic substance ▶ Small cuts on the face, made when shaving, may be treated with a styptic pencil.

sub- comb. form : beneath, below

subalpine [ˌsʌbˈælpaɪn] a. (Bot.) (of plants) which grow below the treeline

subantarctic [ˌsʌbænˈtɑktɪk] a. (Geog.) occurring in latitudes immediately north of the Antarctic circle

subaquatic [ˌsʌbəˈkwɒtɪk] a. (Bot.) (of plants) which grow under or partly under water

subarctic [ˌsʌbˈɑktɪk] a. (Geog.) occurring in latitudes immediately south of the Arctic circle

subatomic [ˌsʌbəˈtɒmɪk] a. (Phys.) (of particles) forming part of an atom ▶ An electron is a subatomic particle.

subclass [ˈsʌbklɑs] n. (Biol.) a subdivision of a class in the taxonomic system ▶ Several previously unknown subclasses of the Musci, or class of mosses, were discovered in the sheltered valleys.

subcutaneous [ˌsʌbkjuˈteɪnɪəs] a. (Med.) having to do with tissue just below the surface of the skin ▶ The injection in the subcutaneous tissue of the arm was quite painless.

subequatorial [ˌsʌbekwəˈtɔrɪəl] a. (Geog.) occurring in latitudes immediately north or south of the equatorial regions

subgenus [ˌsʌbˈdʒinəs] n. (Bot.) a taxonomic division between genus and species

sublimation [ˌsʌblɪˈmeɪʃ(ə)n] n. (Chem.) the process by which a solid, when heated, changes to a gas without going through an

intermediate stage of being a liquid
sublittoral [sʌbˈlɪtər(ə)l] *a. (Biol.)* (of marine organisms) living near the seashore
submontane [sʌbˈmɒnteɪn] *a. (Geog.)* having to do with the lower slopes of a mountain or range of mountains
subsonic [sʌbˈsɒnɪk] *a. (Phys.)* moving at less than the speed of sound ▶ Over highly-populated areas, aircraft are obliged to fly at subsonic speeds.
substance [ˈsʌbstəns] *n. (Chem.)* any material with an identifiable chemical composition
substation [ˈsʌbˌsteɪʃ(ə)n] *n. (Elec.)* a point at which electrical current is received from other power stations to be converted to a different strength or a different type ▶ At the next substation the current was converted from direct to alternating.
subtropical [sʌbˈtrɒpɪk(ə)l] *a. (Bot.)* (of plants) growing in the areas adjacent to the tropics of Cancer or Capricorn ▶ In the subtropical climate of the Pacific islands a number of exotic plants and fruits were found.
sulph- *comb. form* containing sulphur
sulpha drug one of the sulphonamide compounds formerly widely used to treat bacterial infections, before the discovery of penicillin and other antibiotics
sulphate [ˈsʌlfeɪt] *n. (Chem.)* one of the salts formed by the action of sulphuric acid
sulphonamide [sʌlˈfɒnəmaɪd] *n. (Chem.)* one of a group of chemical compounds, including sulpha drugs
sulphur [ˈsʌlfə] *n. (Chem.)* a yellow, non-metallic element found especially in volcanic areas and used in the manufacture of a number of products, including sulphuric acid
sulphuric acid [sʌlˌfjʊərɪk ˈæsɪd] a corrosive liquid made up of sulphur trioxide and water ▶ Sulphuric acid is used in numerous industrial processes, including the manufacture of electric batteries.
sump [sʌmp] *n. (Mech.)* a reservoir at the bottom of a petrol engine in which excess oil etc. is collected for removal
super- *comb. form* : over, above

supercharge [ˈsjupəˌtʃɑdʒ] *v.t. (Mech.)* to boost the power of a petrol engine by increasing the pressure of intake of fuel at high speeds
supercharger [ˈsjupəˌtʃɑdʒə] *n. (Mech.)* the device by which the power of an engine is boosted ▶ The car was fitted with a powerful supercharger which made it roar as it sped along the track.
superconductor [ˈsjupəkənˌdʌktə] *n. (Elec.)* a material that has very low resistance to electric current at low temperatures ▶ Superconductors are used in transistor radios.
supercool [ˈsjupəˌkul] *v.i.* and *v.t. (Phys.)* (of a liquid) to cool or be cooled without freezing to temperatures below those at which freezing normally takes place
superficial [ˌsjupəˈfɪʃ(ə)l] *a.* having to do with the surface, not penetrating deeply
superheat [ˈsjupəˌhit] *v.i.* and *v.t. (Phys.)* (of a liquid) to heat or be heated without boiling to a temperature above that at which it normally boils
supersaturated [ˌsjupəˈsætʃəreɪtɪd] *a. (Chem.)* (of a solution or vapour) containing more of the substance being dissolved or vaporized than is normally possible ▶ Supersaturated fats are considered unhealthy if consumed in quantity.
superscript [ˈsjupəˌskrɪpt] *n.* letters or figures written above the normal line ▶ In the expression x^{23}, the ^{23} is given in superscript.
supersonic [ˌsjupəˈsɒnɪk] *a. (Phys.)* above the speed of sound
suppress [səˈpres] *v.t. (Elec.)* to reduce unnecessary oscillations in a electrical circuit
suppressor [səˈpresə] *n. (Elec.)* a capacitor inserted in an electrical appliance to stop interference with signals being received by a radio or television ▶ Without suppressors, it would be impossible to use a radio or television at the same time as another appliance.
surge [sɜdʒ] *n. (Elec.)* a sudden increase in an electric current ▶ The surge created when

suspend [sə'spend] *v.t.* (*Chem.*) to bring particles of a substance into a state of suspension in a liquid

suspension [sə'spenʃ(ə)n] *n.* (*Chem.*) a state in which particles are supported in a liquid by their own buoyancy

suture ['sutʃə] *n.* (*Med.*) a nylon or wire thread used to stitch a wound after surgery ▶ After five days the wound was seen to be healing well and the sutures were removed.

swept-wing ['swept₁wɪŋ] (of an aircraft) having wings sloping steeply backwards to form a delta-shape ▶ Most people would recognize the swept-wing profile of Concorde.

swing-wing ['swɪŋ₁wɪŋ] (of an aircraft) having wings capable of becoming more swept-back at high speeds

switch¹ [swɪtʃ] *n.* (*Elec.: Mech.*) a device for turning on or off a flow, e.g. of electric current, by completing or breaking the circuit

switch² *v.t.* (*Elec.*) to turn on or off the flow of an electric current by opening or closing a circuit ▶ Don't forget to switch off the lights when you leave.

swivel¹ ['swɪv(ə)l] *n.* (*Mech.*) a device which links a freely-moving part of a mechanism with another part

swivel² *v.i.* (*Mech.*) (of a part of a mechanism) to change position in relation to another part by movement about a linked point ▶ The long trailer swivelled gradually to an acute angle as the articulated lorry turned the steep bend in the road.

symbiosis [₁sɪmbɪ'əʊsɪs] *n.* (*Biol.*) the ability or state of living in close association with something else ▶ At first the wild and newly released animals lived side by side in a state of uneasy symbiosis.

symbol ['sɪmb(ə)l] *n.* (*Maths.*) a character, letter or special mark taken as representing some quantity, etc. ▶ The symbols x and y are frequently used to represent unknown quantities in algebraic formulae.

symmetrical [sɪ'metrɪk(ə)l] *a.* (*Maths.*) (of a figure) which can be divided into two halves, each of which is a mirror-image of the other ▶ The human face is not entirely symmetrical.

symmetry ['sɪmətrɪ] *n.* (*Maths.*) the quality of being symmetrical

symptom ['sɪm(p)təm] *n.* (*Med.*) a physical sign which indicates the presence of disease ▶ The patient presented all the typical symptoms of influenza.

symptomatic [₁sɪm(p)tə'mætɪk] *a.* (*Med.*) typical of or peculiar to an unhealthy condition ▶ The hard lumps in the patient's abdomen were thought to be symptomatic of kidney disease.

syn- *comb. form* together with, resembling

synchronize ['sɪŋkrənaɪz] *v.t.* (*Mech.*) to ensure that something happens in time with something else ▶ Before setting out, the members of the team synchronized their watches.

syndrome ['sɪndrəʊm] *n.* (*Med.*) a combination of symptoms and physical signs which are found together in certain kinds of disease ▶ Acquired Immune Deficiency Syndrome (AIDS) has so far defied all attempts at a cure.

synthetic [sɪn'θetɪk] *a.* (*Chem.*) man-made ▶ Traditional materials, such as wool and cotton, are increasingly being replaced by such synthetic materials as nylon.

syringe [sɪ'rɪndʒ] *n.* (*Med.*) an instrument for injecting fluids ▶ The use of dirty syringes by drug addicts is one of the major causes of the spread of AIDS.

system ['sɪstəm] *n.* a combination of entities which work together to achieve a common purpose ▶ Astronomers are gradually increasing our understanding of the origin and development of the solar system. ▶ The new computer system enables us to access data much more quickly.

systematic [₁sɪstə'mætɪk] *a.* following a coherent and communicable plan, in which all the elements work together ▶ The team of forensic scientists conducted a systematic analysis of all the available evidence before reaching an agreed conclusion.

systemic [sɪsˌtemɪk] *a. (Biol.: Hort.)* (of a weedkiller etc.) which strikes at the core system of a weed, as opposed to killing off its leaves or other external and visible parts ▶ Using a systemic weedkiller has a more fundamental effect by striking at the root system of the plant.

systems analysis analysis, often computerized, of the way in which the object of analysis works, so that a systematic plan of development or improvement may be employed ▶ Out-of-date commercial or industrial enterprises may benefit from the adoption of a systems analysis approach to planning future development, probably involving a computer-generated model for reshaping or revitalizing their operations.

t

tabulate ['tæbjʊleɪt] *v.t.* *(Maths.)* to arrange (data) in tables for easy reference

tail [teɪl] *n.* *(Astron.)* a stream of gas and particles visible in the sky after the passage of a comet ▶ The fiery tail of Halley's comet was clearly visible in the night sky.

tailgate ['teɪlˌgeɪt] *n.* *(Build.)* the gate at the lower end of a lock that controls the passage of water

tailplane ['teɪlˌpleɪn] *n.* *(Aer.)* the horizontal plane at the tail of an aircraft that controls stability in level flight

tailskid ['teɪlˌskɪd] *n.* *(Aer.)* the runner underneath the tail of a small plane, replacing a fixed wheel

tailwind ['teɪlˌwɪnd] *n.* *(Aer.: Meteor.)* a wind moving in the same direction as a vehicle is travelling ▶ Helped by a strong tailwind, the aircraft arrived well ahead of schedule.

talkback ['tɔkˌbæk] *n.* *(Radio)* a system of radio communication between producer and performer during a live radio or television presentation

talkdown ['tɔkˌdaʊn] *n.* *(Radio)* a system of radio communication between ground control and the crew of an aircraft to assist it in landing

tangent ['tændʒənt] *n.* *(Maths.)* a straight line that touches the circumference of a circle without crossing it ▶ The missile glanced off the circular roof and went off at a tangent.

tannin ['tænɪn] *n.* *(Chem.)* a yellowish-brown compound found in certain plants ▶ The sides of the mug were stained with tannin from the strong tea.

tar [tɑ] *n.* *(Chem.)* a thick, black, oily liquid formed from organic matter ▶ Coal tar is used in surfacing roads.

taxonomic [ˌtæksəˈnɒmɪk] *a.* *(Biol.)* having to do with taxonomy ▶ The detailed taxonomic classification of known plants enables newly discovered species to be related to those already included.

taxonomy [tækˈsɒnəmɪ] *n.* *(Biol.)* a system of classifying plants and animals by their origin and structure ▶ The taxonomy includes the categories of class, order, family, genus and species; e.g. the lion is: Class *mammalia*, Order *Carnivora*, Family *Felidae*, Genus and Species *Panthera leo*

technological [ˌtɛknəˈɒdʒɪk(ə)l] *a.* having to do with technology

technology [tɛkˈnɒlədʒɪ] *n.* the practical application of the findings of science, especially to industrial processes ▶ Improvements in technology have increased industrial productivity.

tectonics [tɛkˈtɒnɪks] *n. sg.* *(Geol.)* the study of the process by which the Earth achieved its structure ▶ The science of tectonics explains how the geological features of the Earth took shape.

tele- *comb. form.* distant

telecast ['tɛlɪkɑst] *v.t.* or *n.* *(Radio)* to televise and make a television broadcast, or the broadcast so made

telecommunication [ˌtɛlɪkəˌmjuːnɪˈkeɪʃ(ə)n] *n.* *(Radio)* communication over a great distance, using radio waves

telecommunications [ˌtɛlɪkəˌmjuːnɪkeɪʃ(ə)nz] *n. sg.* *(Radio)* systematized telecommunication over long distances, or the study of such systems ▶ Modern telecommunications systems can flash messages almost instantaneously all over the world and into space.

telegram ['tɛlɪgræm] *n.* *(Radio)* a message sent by telegraph

telegraph ['tɛlɪgrɑf] *n.* *(Radio)* a system or

apparatus for transmitting messages by radio along fixed lines ▶ The telegraph operator relayed warning of the rail crash to stations further along the line.

telemetry [tə'lemətrɪ] *n. (Radio)* a system of transmitting data received by measuring instruments

teleprinter ['telɪˌprɪntə] *n. (Mech.: Radio)* a telegraphic device with a keyboard for converting messages into radio signals for transmission, together with a facility for receiving incoming messages and printing them out ▶ The message coming over the teleprinter caused a great stir in the city.

telescope ['telɪskəʊp] *n. (Phys.)* an optical instrument for increasing the apparent size of distant objects ▶ The approaching vessels could be observed through a telescope, but were invisible to the naked eye.

telescopic [telɪ'skɒpɪk] *a. (Phys.)* having to do with a telescope or the facility afforded by it.

telex ['teleks] *n. (Radio)* the international telegraphic service for transmission and reception of messages on teleprinters which may be hired by individual subscribers ▶ The telex system is now being superseded by fax.

temperature^1 ['temprətʃə] *n. (Phys.)* the degree of heat or cold, e.g. in the atmosphere or in water ▶ The change in the behaviour of semiconductors at different temperatures is a vital feature.

temperature^2 *n. coll. (Med.)* a fever, with a body temperature higher than normal ▶ He complained of a headache and a temperature, and she decided that he had contracted flu.

tensile ['tensaɪl] *a. (Phys.: Metall.)* (of metal etc.) capable of being stretched

tensile strength the ability of a metal or other material to stretch, expressed as the maximum stress it can take without breaking ▶ The tensile strength of the new aluminium alloy was rigorously tested before it could be approved for use.

tension^1 ['tenʃ(ə)n] *n. (Phys.)* the force that tends to stretch a rope, wire etc. ▶ The tension on the hawser increased as the vessel swung in the tide.

tension^2 *n. (Elec.)* the voltage of an electric current ▶ The current was carried by overhead high-tension cables, mounted on pylons.

terminal^1 ['tɜmɪn(ə)l] *n. (Elec.)* the point at which an electric current enters or leaves a battery ▶ Be careful to attach the leads to the correct terminals, positive and negative.

terminal^2 *n. (Comput.)* a point in a network at which a VDU is connected ▶ The new office block has a terminal at every desk.

terminal velocity the highest velocity achieved by a missile or falling body ▶ The terminal velocity of the missile was established as several times the speed of sound.

ternary ['tɜnərɪ] *a. (Maths.)* (of a system of calculation) having a base of 3

tetrahedron [ˌtetrə'hidrən] *n. (Maths.)* a solid figure with four flat planes

therapeutic [ˌθerə'pjutɪk] *a. (Med.)* having to do with the treatment or cure of diseases

therapeutics [ˌθerə'pjutɪks] *n. sg. (Med.)* the branch of medicine concerned with treatment or cure

therapist ['θerəpɪst] *n. (Med.)* someone skilled in or practising therapy

therapy ['θerəpɪ] *n. (Med.)* the treatment or cure of disease

therm [θɜm] *n. (Phys.)* a unit of heat equal to 100,000 British thermal units

thermal ['θɜm(ə)l] *a. (Phys.)* having to do with heat

thermal conductivity [ˌθɜm(ə)l kɒndʌk'tɪvətɪ] a measure of the ability of a substance to convey heat

thermal efficiency [ˌθɜm(ə)l ɪ'fɪʃ(ə)nsɪ] the ratio between the amount of heat supplied to a machine and the amount of work done by it

thermal reactor [ˌθɜm(ə)l ri'æktə] a nuclear reactor driven chiefly by heat

thermal unit [ˌθɜm(ə)l 'junɪt] the amount of heat required to raise 1 lb of water at maximum density through 1°F

thermo- *comb. form* heat

thermochemistry [ˌθɜːməʊˈkemɪstrɪ] n. (Chem.) the branch of chemistry concerned with the amount of heat exchanged during a process of chemical change

thermodynamics [ˌθɜːməʊdaɪˈnæmɪks] n. sg. (Phys.) the branch of physics concerned with the relationship between heat and other forms of energy

thermoelectric [ˈθɜːməʊɪˈlektrɪk] a. (Phys.: Elec.) having to do with the conversion of heat into electrical energy

thermonuclear [ˌθɜːməʊˈnjuːklɪə] a. (Phys.) having to do with nuclear fission occurring at very high temperature ▶ She said that the continued existence of thermonuclear weapons was a threat to the survival of the human race.

thrombosis [θrɒmˈbəʊsɪs] n. (Med.) a medical condition in which a thrombus is present ▶ Coronary thrombosis is a common cause of death.

thrombus [ˈθrɒmbəs] n. (Med.) a clot in the heart or a blood vessel, impeding the flow of blood

throttle [ˈθrɒtl] n. (Mech.) the mechanism that controls the flow of fuel in a petrol engine ▶ The driver opened the throttle wide and the car roared down the track.

thrust [θrʌst] n. (Mech.) the motive force in a forward direction generated by a propeller or a jet engine ▶ The tremendous increase in thrust provided by jet propulsion has made supersonic flight a reality.

time zone [ˈtaɪm ˌzəʊn] a region in which the same standard time is observed ▶ A flight across the United States involves going through several time zones.

tissue [ˈtɪsjuː] n. (Biol.: Med.) any part of a living organism comprising many cells which have the same structure and function ▶ Surgeons can now transfer live tissue from one part of the body to another.

titration [tɪˈtreɪʃ(ə)n] n. (Chem.) the measurement, e.g. of the concentration of a liquid, made by adding it to another liquid of known concentration until a chemical reaction between the two is completed ▶ Titration is used in various industrial processes.

torque [tɔːk] n. (Phys.: Eng.) the force which causes rotation ▶ As the plane touched down, the pilot used the rudder to counteract the torque of the propeller and avoid veering off the runway.

toxic [ˈtɒksɪk] a. (Chem.: Med.) having to do with poison, poisonous

toxic waste poisonous effluent or other unwanted matter, usually left after an industrial process has been completed ▶ Disposing of such toxic waste as radioactive materials is an increasing problem in the industrialized countries.

toxin [ˈtɒksɪn] n. (Chem.: Med.) a poisonous substance ▶ The snake injected a deadly toxin into the bloodstream of its prey.

trans- comb. form movement or change across, or from one thing to another

transceiver [trænˈsiːvə] n. (Radio) a radio set capable of both transmitting and receiving signals

transcribe [ˌtrænˈskraɪb] v.t. (Comput.) to copy data from one form of storage to another ▶ The material was transcribed from punched cards to floppy disks, making it less bulky and easier to access.

transducer [ˌtrænzˈdjuːsə] n. (Phys.: Mech.) a device for converting one form of energy to another ▶ Everyday types of transducer include the telephone, microphone, and loudspeaker.

transform [trænsˈfɔːm] v.t. (Maths.) to change a fraction from one form to another ▶ The fraction $\frac{1}{2}$ can be transformed to 0·5.

transformer [trænsˈfɔːmə] n. (Elec.) a device for changing the voltage of an alternating current passing from one circuit to another ▶ Many electrical appliances made in one country can be used in other countries only by adding a transformer to the circuit.

transfusion [trænsˈfjuːʒ(ə)n] n. (Med.) the process or result of making an injection of blood or plasma from one person into the veins of another ▶ The patient had lost so much blood that a transfusion was

urgently necessary.
transistor [trænˈzɪstə] *n. (Elec.)* a small electronic device by which a semiconductor replaces a vacuum tube valve in controlling the electrical current in a circuit
transistor radio [trænˌzɪstə ˈreɪdɪəʊ] a portable radio in which transistors take the place of valves ▶ Among the advantages of transistor radios are that they are battery-operated and can be compact enough to fit into small spaces. ▶ The disadvantages of transistor radios are the same.
translucent [trænsˈlus(ə)nt] *a. (Phys.)* capable of allowing light to pass through ▶ For the windows of the spacecraft it was necessary to develop a material that was tough and durable, but efficiently translucent.
transmission¹ [trænsˈmɪʃ(ə)n] *n. (Mech.)* a mechanism which transmits power from one part of a machine to another ▶ The power generated by the engine of a car is passed to the axles by the transmission mechanism.
transmission² *n. (Radio)* the process or act of transmitting a radio signal, or the signal transmitted ▶ The daily transmissions by the local radio station were extremely popular.
transmit [trænsˈmɪt] *v.t. (Radio)* to send out radio signals
transmitter [trænsˈmɪtə] *n. (Radio)* an apparatus for sending out radio signals
transplant [trænsˈplɑnt] *n.* or *v.t. (Med.)* to replace a diseased or damaged organ with one taken from another body; the organ which is transplanted, or the surgical operation involved ▶ Though the technique of transplanting organs has only recently been developed, heart transplants are now becoming quite common.
transpose [trænsˈpəʊz] *v.t. (Maths.)* to switch a value from one side of an equation to another, changing from + to −, or from − to +
transverse [trænsˈvɜs] *a. (Mech.)* (of an engine) placed parallel to the axle, instead of at right angles to it
treatment¹ [ˈtritmənt] *n. (Med.)* any medical procedure designed to cure or alleviate
treatment² *n. (Chem.)* the process of being changed from one state to another by chemical or bacteriological means ▶ The sewage is conveyed to a sewage farm for treatment before being recycled as agricultural fertilizer.
trigonometry [ˌtrɪgəˈnɒmətrɪ] *n. (Maths.)* the branch of mathematics dealing with the relationships of the sides and angles of triangles and applying these to other fields, including astronomy.
trim [trɪm] *v.t. (Aer.: Naut.)* to adjust the balance of a vessel or aircraft by redistributing the load or altering the setting of the controls ▶ The sails had often to be trimmed to suit the changes in the direction of the wind.
tripod [ˈtraɪpɒd] *n.* a three-legged stand, stool, etc. ▶ The photographer needed to mount his camera on a tripod in order to keep it absolutely still.
trisect [traɪˈsekt] *v.t. (Maths.)* to divide into three parts, usually equal
tropic [ˈtrɒpɪk] *n. (Geog.)* one of the two lines of latitude situated at 23°27′ from the equator ▶ The tropic of Cancer lies north of the equator, and the tropic of Capricorn to the south.
tropical [ˈtrɒpɪk(ə)l] *a. (Geog.: Meteor.)* having to do with or lying between the tropics, where the weather is very hot ▶ Vegetation in the tropical rainforests is very rich and provides a natural habitat for hundreds of species of animals and plants, especially trees.
troposphere [ˈtrɒpəsfɪə] *n. (Phys.)* the lowest level of the Earth's atmosphere
trough¹ [trɒf] *n. (Meteor.)* an area of low pressure between two areas of higher pressure ▶ The whole country lay in the trough of a depression for most of the summer.
trough² *n. (Phys.)* (of a wave of sound, light, etc.) the lowest point
trunk [trʌŋk] *n. (Biol.)* the main part of the body
trunk line a direct link between two telephone installations some distance apart

tumbler [ˈtʌmblə] *n. (Mech.)* that part of a lock that is moved by a key

tumour [ˈtjumə] *n. (Med.)* a swelling, often malignant, on a part of the body ▶ She died of a brain tumour at a very early age.

tune¹ [tjun] *v.t. (Radio)* to adjust a receiver in order to obtain the correct signal ▶ It was difficult to tune the set because of interference from other stations.

tune² *v.t. (Mech.)* (of an engine) to adjust the fuel supply etc. for maximum efficiency ▶ The engine of the racing car was finely tuned for high speeds.

turbine [ˈtɜbaɪn] *n. (Eng.)* an engine driven by a wheel which is turned by the action of a fluid or a gas ▶ Turbines driven by water from the dam generate electrical energy.

turbo- *comb. form* powered by a turbine

turbo-electric (of an electrical generator) driven by a turbine

u

udometer [juˈdɒmɪtə] *n.* (*Meteor.*) a rain-gauge

UHF (*abbr.*) ultra-high frequency

ulcer [ˈʌlsə] *n.* (*Med.*) an open sore on a part of the body, usually with pus or a discharge

ulcerate [ˈʌlsəreɪt] *v.i.* (*Med.*) to develop ulcers ▶ The child's tongue was ulcerated and sore.

ultra- *comb. form* beyond, on the other side of

ultra-high frequency [ˌʌltrəˌhaɪ ˈfrikwənsɪ] (*Radio*) radio frequencies between 300 and 3,000 megahertz

ultra-short waves [ˌʌltrəˌʃɔt ˈweɪvz] (*Radio*) radio waves below 10 metres in wavelength

ultrasonic [ˌʌltrəˈsɒnɪk] *a.* (*Phys.*) having to do with or using sound waves of higher than audible frequency

ultrasound [ˈʌltrəˌsaʊnd] *n.* (*Phys.: Med.*) ultrasonic waves used especially for medical diagnosis

ultraviolet [ˌʌltrəˈvaɪələt] *a.* (*Phys.*) (of rays) having a wavelength shorter than the violet end of the visible spectrum but longer than X-rays

umbilical [ʌmˈbɪlɪk(ə)l] *a.* (*Med.*) having to do with the umbilicus

umbilical cord the cord-like structure of vessels connecting the foetus with the placenta

umbilicus [ʌmˈbɪlɪkəs] *n.* (*Med.*) the navel

uni- *comb. form* one, single

unicellular [juːnɪˈseljʊlə] *a.* (*Biol.*) consisting of only one cell ▶ Some forms of algae are unicellular.

unilateral [juːnɪˈlætər(ə)l] *a.* (*Bot.*) (of a plant) having certain features all on one side

unipolar [juːnɪˈpəʊlə] *a.* (*Elec.: Med.*) showing only one kind of polarity; (of nerve cells) having only one pole

unit [ˈjuːnɪt] *n. a* single person or item, or a group of persons or items considered as one for the purposes of calculation etc. ▶ Each individual forms a separate unit. ▶ The items were organized in units of five. ▶ The unit of work done is the erg.

universal [juːnɪˈvɜːs(ə)l] *a.* having to do with the universe; taking all things together

universal adaptor (*Elec.*) a device that allows any kind of electric plug to be used in any kind of socket

universal joint (*Mech.*) a device that joins two things while allowing movement in all directions

universe [ˈjuːnɪvɜːs] *n.* all existing things taken together, the cosmos

unleaded [ʌnˈledɪd] *a.* (*Chem.*) (of petrol) not having an admixture of lead

unrefined [ʌnrɪˈfaɪnd] *a.* (*Chem.*) (of sugar, petroleum, etc.) not processed into a pure form

uranium [jʊˈreɪnɪəm] *n.* (*Chem.: Phys.*) a white, metallic element found in several minerals which is radioactive and can be used in producing nuclear fission

uranium bomb an atom bomb which uses uranium as an explosive

urea [ˈjʊərɪə] *n.* (*Chem.: Agric.*) a chemical compound found in urine and used as a fertilizer

urethra [jʊəˈriːθrə] *n.* (*Med.*) the duct through which urine is discharged from the bladder

urinate [ˈjʊərɪneɪt] *v.i.* (*Med.*) to expel waste fluid from the body via the urethra

urine [ˈjʊərɪn] *n.* (*Biol.*) waste fluid secreted by the kidneys, collected in the bladder, and expelled via the urethra

uterus [ˈjuːtərəs] *n.* (*Med.*) the womb

V

vaccinate [ˈvæksɪneɪt] *v.t.* (*Med.*) to inoculate with vaccine ▶ All the children in the village school were vaccinated against smallpox.

vaccination [ˌvæksɪˈneɪʃ(ə)n] *n.* (*Med.*) the process or result of being vaccinated

vaccine [ˈvæksin] *n.* (*Med.*) a liquid in which specially treated micro-organisms of a disease are suspended for use in inoculation against that disease ▶ Fresh supplies of vaccine were flown to the infected area in an attempt to limit the spread of the disease.

vacuum [ˈvækjʊəm] *n.* (*Phys.*) a space from which all air has been evacuated ▶ Living creatures cannot survive in a vacuum.

vacuum cleaner [ˈvækjʊəm ˌklinə] an apparatus which cleans by creating a vacuum into which dust etc. is sucked

vacuum flask [ˈvækjʊəm ˌflɑsk] a flask having double walls with a vacuum between them, which acts as insulation and slows down any change in temperature of the contents ▶ The vaccine ampoules were kept cool in a vacuum flask of iced water.

valency [ˈveɪlənsɪ] *n.* (*Chem.*) the combining power of atoms measured by the number of hydrogen atoms they can displace in forming chemical compounds

value [ˈvælju] *n.* (*Maths.*) a particular number or size ▶ In solving the equation, the value of x was taken as zero.

valve¹ [vælv] *n.* (*Mech.*) a device that controls the flow of a liquid or a gas ▶ Turning the tap opened the valve and allowed more water to enter the tank.

valve² *n.* (*Biol.*) the shell and membrane of a certain kind of shellfish ▶ A mollusc has a shell which consists of two separate pieces and is therefore a bivalve.

valve³ *n.* (*Radio*) a vacuum tube which contains electrodes and controls the flow of electric current ▶ Sensitive and fragile valves are now being replaced by transistors.

vane [veɪn] *n.* (*Mech.*) a fin or blade, e.g. of a propeller or fan

vaporize [ˈveɪpəraɪz] *v.t.* and *v.i.* (*Phys.*) to turn into or become vapour

vapour [ˈveɪpə] *n.* (*Phys.*: *Chem.*) particles of moisture etc. visible as clouds in the air

vapour trail a trail of vapour left by an aircraft flying through cold air

variable [ˈveərɪəbl] *n.* (*Maths.*) a quantity or number that can be given a range of different values ▶ The equation was difficult to solve because there were so many variables.

vascular [ˈvæskjʊlə] *a.* (*Med.*) having to do with or composed of vessels that convey fluid

vasectomy [vəˈsektəmɪ] *n.* (*Med.*) an operation performed on men to make them sterile

VDU [ˌvidiˈju] (*abbr.*) (*Comput.*) visual display unit – the screen of a computer or word-processor

vector [ˈvektə] *n.* (*Maths.*) a line showing the time and direction of a given quantity in space or on a diagram

veer [vɪə] *v.i.* (*Naut.*: *Meteor.*) (of the wind) to diverge from an original path in a clockwise direction north of the equator and in an anticlockwise direction south of the equator ▶ As the ship neared the area of the cyclone, the winds veered sharply and became stronger.

vegetable [ˈvedʒ(ɪ)təbl] *a.* (*Bot.*) having to do with plants as opposed to minerals or animals

vegetation [ˌvedʒɪˈteɪʃ(ə)n] *n.* (*Bot.*) plant life

vein [veɪn] *n.* (*Biol.: Med.*) a major blood vessel, which conveys blood to the heart

velocity [vɪˈlɒsətɪ] *n.* (*Phys.*) the rate at which a body changes its position ▶ Falling objects increase their velocity at a rate of 32 feet per sec. per sec.

venereal [vəˈnɪərɪəl] *a.* (*Biol.: Med.*) having to do with or resulting from sexual intercourse

venereal disease [vəˈnɪərɪəl dɪˌziz] a disease transmitted via sexual intercourse ▶ Condoms are worn as protection against venereal disease.

venom [ˈvenəm] *n.* (*Med.*) poison of the sort found in the fangs of snakes and other reptiles ▶ As the venom from the cobra's bite entered his bloodstream, he lost consciousness.

venomous [ˈvenəməs] *a.* (*Biol.*) (of snakes, spiders, etc.) having bites that transmit venom

venous [ˈvinəs] *a.* (*Biol.: Med.*) having to do with veins

vent [vent] *n.* (*Mech.*) an aperture for the passage of air or a gas ▶ Excess fumes escaped from the chamber through a vent in the roof.

ventilate [ˈventɪleɪt] *v.t.* (*Mech.*) to ensure the passage of air through a building etc. by the use of fans and vents ▶ The mineshaft was ventilated by a number of huge fans.

ventilator[1] [ˈventɪleɪtə] *n.* (*Mech.*) a device or fitting used in ventilation of a room etc.

ventilator[2] *n.* (*Med.*) a mechanical means of assisting breathing ▶ He was in a deep coma and was put on a ventilator to make it possible for him to breathe.

ventral [ˈventr(ə)l] *a.* (*Med.*) ▶ having to do with the abdomen or belly ▶ His protruding stomach was due to ventral hernia.

ventricle [ˈventrɪk(ə)l] *n.* (*Med.*) a hollow chamber within an organ of the body ▶ Tests revealed poor blood circulation in the left ventricle of her heart.

vermicide [ˈvɜmɪsaɪd] *n.* (*Chem.*) a medicine or drug that kills worms

vernal [ˈvɜn(ə)l] *a.* (*Astron.*) having to do with or related to Spring

vernal equinox [ˌvɜn(ə)l ˈikwɪnɒks] the date in Spring when night and day are of equal length

verso [ˈvɜsəʊ] *n.* the reverse side of a sheet of printed paper ▶ The conditions of sale are printed on the verso of the invoice.

vertebra [ˈvɜtɪbrə] *pl.* **vertebrae** [ˈvɜtɪbri] *n.* (*Biol.*) one of a number of bones which form the spine

vertebral [ˈvɜtɪbr(ə)l] *a.* (*Biol.*) having to do with the spine

vertebral column the spine

vertebrate [ˈvɜtɪbrət] *n.* or *a.* (*Biol.*) (of an animal) having a spine ▶ Man is a vertebrate.

vertical [ˈvɜtɪk(ə)l] *a.* or *n.* (*Maths.*) (of a line) at right angles to the horizontal ▶ Vertical take-off aircraft do not need a long runway.

vesicle[1] [ˈvesɪk(ə)l] *n.* (*Med.*) a small cavity or sac in the body which contains liquid ▶ Small, superficial vesicles are commonly called blisters.

vesicle[2] *n.* (*Bot.*) a bladder-like part of a plant, such as occurs in some kinds of seaweed

vesicle[3] *n.* (*Geol.*) a round cavity in a rock formation caused by the expansion of gas when the rock cooled

vessel[1] [ˈvesl] *n.* (*Med.*) a tube-shaped element in the body that conveys fluids such as blood ▶ The main blood vessels are the arteries and the veins.

vessel[2] *n.* a small dish or other receptacle for holding liquids or other substances

vibrate [vaɪˈbreɪt] *v.i.* (*Mech.*) to move rapidly to and fro with a slight shaking movement

vibration [vaɪˈbreɪʃ(ə)n] *n.* (*Phys.*) the act of moving to and fro, and the sound so created

vibrator [vaɪˈbreɪtə] *n.* (*Elec.*) an electrical device that produces controlled vibrations ▶ Vibrators are commonly used in massage.

video- [ˈvɪdɪəʊ] *comb. form* concerned with the creation of a visual image by television

video cassette a cassette containing magnetic

tape for recording visual images
video frequency the electrical frequency on which a given video signal is transmitted or received ▶ Video recorders have to be tuned to the correct video frequency.
video (cassette) recorder a machine for recording and playing back television programmes on video cassettes
video (tape) recording a recording of a television programme on magnetic tape for later transmission
vinyl ['vaɪnɪl] *n. (Chem.)* one of a number of plastic substances containing vinyl resin ▶ A common vinyl product used to insulate electrical leads is PVC.
viral ['vaɪ(ə)r(ə)l] *a. (Med.)* concerned with or caused by a virus ▶ A number of viral infections have only recently been identified as such.
virulent ['vɪrʊlənt] *a. (Med.)* extremely poisonous; (of a disease) very infectious or very violent in its effect
virus[1] ['vaɪ(ə)rəs] *n. (Med.)* one of a group of tiny entities which live and multiply inside the cells of the body and may be the cause of a number of diseases ▶ Researchers have now isolated several strains of virus that cause HIV and may lead to AIDS.
virus[2] *n. (Comput.)* a self-replicating computer program which damages or destroys the memory or other programs in a computer ▶ The entire year's records were lost when the company's computer was invaded by a virus.
viscera ['vɪsərə] *n. pl. (Med.)* the internal organs of the main parts of the body, especially the abdomen
viscosity [vɪ'skɒsətɪ] *n. (Phys.)* the state or degree of being viscous
viscous ['vɪskəs] *a. (Phys.)* (of a fluid) thick and sticky ▶ The fluid was so viscous that it could hardly be poured from the jar.
vitamin ['vɪtəmɪn] *n. (Med.)* one of a group of substances, usually denoted by letters, which are essential to the healthy functioning of the human body ▶ Lime juice contains vitamin C.
vitriol ['vɪtrɪəl] *n. (Chem.)* a popular name for sulphuric acid
vivi- *comb. form* live
viviparous [vɪ'vɪpərəs] *a. (Biol.)* bringing forth live young ▶ Human beings are viviparous, but creatures that lay eggs are not.
vivisection ['vɪvɪˌsekʃ(ə)n] *n. (Biol.)* the carrying out of experiments on live animals in the interests of science ▶ Many people oppose the practice of vivisection for commercial gain.
volatile ['vɒlətaɪl] *a. (Chem.)* (of a liquid) capable of turning rapidly into a vapour ▶ High-grade petrol is more volatile than diesel oil and ignites more easily.
volt [vəʊlt] *n. (Elec.: Meas.)* a measure of the potential of an electric current ▶ 1 volt is defined as the difference in potential between two points on a conductor carrying a current of 1 ampere, when the current dissipated between them is 1 watt. ▶ High-tension cables carry loads of thousands of volts.
voltage ['vəʊltɪdʒ] *n. (Elec.)* electric potential, expressed in terms of volts ▶ Care must be taken to ensure that batteries used are of the correct voltage.
volume ['vɒljum] *n. (Phys.)* the mass of a substance contained in a given space ▶ The volume of water in the tank was not sufficient to put out the fire. ▶ An object totally immersed in water will displace its own volume, but a floating object will displace its own weight.
vortex ['vɔteks] *pl.* **vortices** ['vɔtɪsiz] *n. (Phys.)* a whirling mass of matter such as gas or liquid ▶ A whirlpool and a whirlwind are both vortices.
vulcanize ['vʌlkənaɪz] *v.t. (Chem.)* to improve the durability of substances such as rubber by treating them with sulphur compounds at great heat or pressure ▶ Vehicle tyres are made of vulcanized rubber.

W

wafer [ˈweɪfə] *n.* (*Elec.*) a thin slice of silicon or other semiconductor which carries a number of circuits and can easily be removed

wake [weɪk] *n.* (*Naut.*) the visible track left in the sea after a ship has passed.

walkie-talkie [ˌwɔkɪˈtɔkɪ] a portable combined radio transmitter and receiver ▶ Policemen with walkie-talkies were controlling the crowds.

wall [wɔl] *n.* (*Biol.*) the outer layer of a bodily organ, such as the stomach

wane [weɪn] *v.i.* (*Astron.*) (of the Moon) to become smaller and less bright

warm [wɔm] *a.* (*Phys.*) having a certain amount of heat

warm-blooded [ˌwɔmˈblʌdɪd] (of animals) having a constant, warm body temperature, usually between 98°F and 112°F (36·6°C and 44·4°C) ▶ Mammals and birds are warm-blooded.

warm front [ˈwɔm ˈfrʌnt] (*Meteor.*) the front edge of an advancing mass of warm air ▶ A warm front moved slowly across the country, bringing a rise in temperature, with rain.

warp [wɔp] *v.i.* (of wood) to lose shape because of heat or damp ▶ Because of the persistent rain the window frames became severely warped and would not close.

wart [wɔt] *n.* (*Med.*) a small, hard growth on the surface of the skin, usually benign

waste¹ [weɪst] *v.i.* (*Biol.: Med.*) to gradually lose health, strength, and weight ▶ As the illness took hold, he began visibly to waste away.

waste² *n.* (*Eng.: Mining*) that part of a body of material that is not utilized during an industrial process and is therefore discarded ▶ Huge tips of waste from the ironworks disfigured the valley. ▶ Ecologists are worried about the dumping of nuclear waste.

watt [wɔt] *n.* (*Elec.*) a unit of electrical power equal to that of a current of 1 ampere over a potential difference of 1 volt

wattage [ˈwɔtɪdʒ] *n.* (*Elec.*) electrical power expressed in terms of watts ▶ The light from the 40-watt bulb was so dim that we could not see to read.

wave [weɪv] *n.* (*Phys.*) an undulating disturbance, resembling on paper the waves of the sea, which conveys such forms of energy as light and sound without altering the physical shape or appearance of the matter through which it passes

waveband [ˈweɪvˌbænd] *n.* (*Radio*) the range of wavelengths necessary for the transmission of a particular type of radio signal ▶ The new station began broadcasting in the medium waveband.

wavelength [ˈweɪvˌleŋθ] *n.* (*Phys.: Radio*) the distance between the crests of successive waves ▶ Modern radios operate on a number of wavelengths, from long-wave to ultra-short.

wax [wæks] *v.i.* (*Astron.*) (of the Moon) to grow bigger and brighter

wedge [wedʒ] *n.* (*Mech.*) a piece of wood or metal, thicker at one end than the other, used for a variety of purposes, such as splitting or raising other objects ▶ A thick wedge had been jammed under the door from the inside, and it was impossible to get in.

weight [weɪt] *n.* (*Phys.*) the force with which bodies tend towards the centre of the earth under the influence of gravity

weightlessness [ˈweɪtləsnəs] *n.* (*Phys.*) the impression of having no weight because the force of gravity is not felt ▶ Astronauts in spacecraft have to learn to operate in conditions of weightlessness, in which unsecured solid objects float in mid-air.

weir [wɪə] *n. (Eng.)* a dam built across a river to raise the level of the water

weld [weld] *v.t. (Metall.)* to join two pieces of metal by heating them and hammering until they unite ▶ A new section had to be welded on to the axle before the vehicle could move.

white [waɪt] *a. (Phys.)* referring to material which rejects light and therefore has no visible colour ▶ Though white is commonly referred to as a colour, it in fact describes the state of seeming to be without one.

white heat heat so intense that the heated substance emits white light

white light light which contains all the visible wavelengths at the same intensity ▶ Sunlight is white light.

winch [wɪntʃ] *n.* or *v.t. (Mech.)* a crank operated manually or power-driven and used for pulling or raising heavy items; to use such a crank ▶ The cargo was winched aboard.

wind [wɪnd] *n. (Meteor.)* air in motion, a strong natural current of air

wind chill factor *see* **chill factor**

wind tunnel [ˈwɪnd ˌtʌn(ə)l] a device for testing models of new aircraft designs by studying their behaviour in a controlled artificial stream of air

windlass [ˈwɪndləs] *n. (Mech.)* another name for winch

wingnut [ˈwɪŋˌnʌt] *n. (Mech.)* a threaded nut with two wing-like projections to facilitate manual tightening or loosening

wireless [ˈwaɪələs] *n. (Radio)* wireless telegraphy – the transmission of sounds through the air by electromagnetic waves, without the use of wires ▶ The term wireless has now been replaced by radio.

word processor [ˈwɜd ˌprəʊsesə] an electronic device, equipped with a VDU, for typing, editing and printing documents ▶ Most business offices are now equipped with word processors rather than typewriters.

work [wɜk] *n. (Phys.)* transfer of energy, expressed in terms of force and distance moved ▶ The unit of work done is the erg.

worm [wɜm] *n. (Mech.)* the spiral part of a screw

worm gear [ˈwɜmˌgɪə] a metal shaft with a spiral groove which meshes with a cogged wheel to transfer motion

wow [waʊ] *n. (Radio)* a distortion of sound which may occur in the lower audio frequencies in a recording system ▶ A common cause of wow is variation in the speed of a turntable.

wrought iron [ˈrɔtˌaɪən] iron which has been made easy to work by having all the impurities removed ▶ The entrance to the factory was marked by tall, wrought-iron gates.

x

xero- *comb. form* dryness

xerography [zɪəˈrɒgrəfɪ] *n. (Photo.: Mech.)* a process of photocopying from an electrostatic image on a plate or cylinder

xerox [ˈzɪərɒks] *v.t.* or *a.* or *n. (Photo.: Mech.)* the process or result of dry photocopying
▶ Xeroxing copies of documents is quicker and easier than retyping, and the copies are clearer than carbons.

X-ray [ˈeksreɪ] *n. (Phys.: Med.)* electromagnetic radiation, especially as a way of photographing internal organs and bones for diagnosis and treatment

X-ray astronomy the detection and measurement of X-rays emitted by bodies in space

X-ray star a star that emits X-rays

X-ray telescope a telescope that detects and measures X-rays emitted by stars

X-ray therapy the use of X-rays in medical treatment of diseases

y

yaw [jɔ] *v.i. (Naut.: Aer.)* (of a ship or aircraft) to deviate from a straight course by swinging about its own axis

yaws [jɔz] *n. sg. (Med.)* an infectious tropical disease, characterized by sores which look like raspberries

yield [jild] *n. (Phys.: Chem.)* the energy released by a nuclear explosion, or the amount of a product obtained by a chemical reaction

Z

zenith [ˈzenɪθ] *n. (Astron.)* the point in space directly above the observer ▶ At midday the Sun reaches its zenith.

zero [ˈzɪərəʊ] *n. (Maths.)* the point on a scale from which positive and negative are reckoned ▶ Zero on the Fahrenheit temperature scale is 32° below the freezing-point of water, which is zero on the centigrade and celsius scales.

zip [zɪp] *(abbr.)* zone improvement plan

zip code the postal code in an American address

zodiac [ˈzəʊdɪæk] *n. (Astron.)* a zone or belt in the sky extending to about 8° on each side of the line that the Sun appears to travel in one year

zone[1] [zəʊn] *n. (Geog.)* one of the five belts into which the Earth is conventionally divided, along lines of latitude ▶ The five zones are the Torrid, North and South Temperate, and North and South Frigid.

zone[2] *n. (Geol.)* a layer of rock etc. which illustrates a historical period the in Earth's development by its fossil content and structure

zone[3] *n. (Biol.)* an area of the world which has common flora and fauna

zoology [zuˈɒlədʒɪ] *n.* the branch of biology that studies the structure, classification etc. of animals

zoological garden [zʊəˈlɒdʒɪk(ə)l ˈgɑd(ə)n] *(abbr.* **zoo**) a public park or garden in which a collection of animals is kept for public viewing

zoom [zum] *v.i. (Phys.: Mech.)* (of a camera) to use a zoom lens to change the apparent size of an object being photographed

zoom in to use a zoom lens to rapidly increase the apparent size of an object being photographed

zoom lens a camera lens that can make an object being photographed seem larger or smaller by changing the focal length

zoom out to use a zoom lens to rapidly decrease the apparent size of an object being photographed